Crystal
Healing

The **SCIENCE** and **PSYCHOLOGY**
Behind *What Works*,
What Doesn't,
and *Why*

DAN R. LYNCH & JULIE A. KIRSCH, PHD

ADVENTURE PUBLICATIONS
CAMBRIDGE, MINNESOTA

ACKNOWLEDGMENTS

Dan R. Lynch and Julie A. Kirsch would like to thank Bob and Nancy Lynch for helping acquire specimens and resources.

DEDICATION

Dedicated to Julie's cousin, Will (Fred) Potthoff, whose unmatched sense of humor and devotion to science is greatly missed but not forgotten.

Cover and book design by Travis Bryant
Edited by Brett Ortler

Cover photos by Dan R. Lynch unless otherwise noted. Front: **Bildagentur Zoonar GmbH:** Crystals (top); **pukach/Shutterstock.com:** Beakers (bottom). Back: **Albert Russ/Shutterstock.com:** Crystals (top left).

All photos by Dan R. Lynch unless otherwise noted. **Meredith Boyter Newlove,** M.S.: pg. 73; **NASA/JPL-Caltech:** pg. 123

Images used under license from Shutterstock.com: **Cagla Acikgoz:** pg. 132; **Antoine2K:** pg. 29; **aregfly:** pg. 159 (top left); **Asya Babushkina:** pg. 155 (top right); **bellabankes:** pg. 42; **Mark Brandon:** pg. 179 (left); **Anastasia Bulanova:** pg. 170; Yut chanthaburi: pg. 167 (bottom left); **Marcel Clemens:** pp. 13, 175 (bottom left); **Alex Coan:** pg. 151 (top left); **Dafinchi:** pg. 28; **Damir B:** pg. 177 (top right); **Delennyk:** pp. 169 (top left), 173 (top right); **DiKiYaqua:** pg. 143 (top right); steve estvanik: pg. 165 (bottom left); **Everett Collection:** pg. 111; **Marco Fine:** pp. 25 (top right), 136; **first vector trend:** pg. 112; **Marlon Fukunaga:** pg 151 (bottom left); **galka3250:** pg. 181 (top right); **Alexander Gold:** pg. 32; **Horiyan:** pg. 135 (top right); **igsin:** pg. 151 (top right); **Nenad Ilic:** pg. 93; **Sebastian Janicki:** pg. 142; **Jirik V:** pp. 149 (top right), 159 (top right); **Kaia-Liisa:** pg. 43; **KrimKate:** pg. 173 (top left); **liewluck:** pg. 179 (right); **Madlen:** pg. 157 (top right); **Matus Madzik:** pg. 137 (top left); **Mahirart:** pg. 164; **Stefan Malloch:** pp. 40, 133 (top right); **Marzolino:** pg. 110; **Holly Mazour:** pp. 34, 38, 87, 97; W. Scott McGill: pg. 151 (bottom right); **Miiisha:** pg. 159 (bottom left); **My Good Images:** pg. 178; **Nordroden:** pg. 133 (bottom left); **Oks-Art:** pg. 165 (top left); **oorka:** pg. 11 (bottom); **Roy Palmer:** pp. 163 (bottom left), 184 (bottom left); **J. Palys:** pp. 141 (top right), 167 (top left & top right); **Julian Popov:** pg. 167 (top right); **July Prokopiv:** pg. 85; **Roy photo:** pg. 84; **Minakryn Ruslan:** pp. 135 (top left), 139 (middle left), 145 (top right), 147 (top left), 149 (top left), 162; **Albert Russ:** pp. 6, 138, 139 (top left), 153 (top left), 167 (bottom right), 174, 181 (top left & bottom left); **Arnav Samant:** pg. 175 (top right); **Jaroslav Sekeres:** pg. 16; **Shark_749:** pg. 128; **Andres Sonne:** pg. 153 (top right); stockcreations: pg. 89; STUDIO492: pg. 139 (bottom left); **Josep Suria:** pg. 83; Teo Tarras: pg. 36; **Rafael Tonani:** pg. 37; TR_Studio: pg. 163 (bottom right); **giedre vaitekune:** pg. 95; vvoe: pp. 96, 137 (top right), 144, 163 (top right), 171 (top left & top right), 177 (top left), 185; wavebreakmedia: pp. 91, 107; **Bjoern Wylezich:** pp. 158, 163 (top left), 166, 172, 175 (top left), 180; Vladimir Ya: pg. 165 (bottom right); **Nikki Zalewski:** pg. 92; **Stellar Gems:** pg. 184 (top right)

Crystal Healing
Copyright © 2021 by Dan R. Lynch and Julie A. Kirsch
Published by Adventure Publications
An imprint of AdventureKEEN
310 Garfield Street South, Cambridge, Minnesota 55008
(800) 678-7006, www.adventurepublications.net
All rights reserved
Printed in the United States of America
ISBN 978-1-59193-917-7 (pbk.); ISBN 978-1-59193-918-4 (ebook)

TABLE OF CONTENTS

INTRODUCTION

Crystal healing is often described as nothing short of a miracle cure, and crystals are used to treat everything from physical pain to emotional concerns, but what is crystal healing? For some, it is merely a curiosity, a way to connect with nature, but for others, it is a replacement for traditional medical care. No matter how far one takes their belief in crystal healing, the core concept is the same: it supposes that everything in the universe is connected by spiritual energy, and that crystals can act as conduits or emitters to connect our minds and bodies to that energy. But does it actually work?

Crystals are undeniably appealing. Their geometry is enigmatic and precise, their colors sometimes unbelievably vibrant, and many of the finest specimens are so pristinely transparent that you could read this page right through them. Crystals are perfectly natural, but many appear so well-formed that they seem to have been constructed; for some, this apparent perfection implies that crystals could contain some sort of otherworldly power. And that's exactly the kind of enthusiastic reverence and awe that crystal-healing authors cultivate, exploit, and sell. But after centuries of research, we know crystals aren't magic; in fact, they are well understood, and they certainly don't impart any "healing energies."
So why do some people still believe that crystals can heal?

Quartz is one of the most important minerals used by crystal healers.

Individuals who feel they have found genuine success with crystal healing often recommend the practice to others, especially on the internet. Many crystal-healing adherents often lead an alternative lifestyle and have little trust in modern medicine. But crystal healing is also popular because many of its proponents stand to make a profit, whether via selling books, hawking crystal specimens, or as therapists performing treatments. These "healers" sell a lifestyle, and some of their followers have even replaced tested, evidence-based treatments with crystal healing and other alternative practices when facing illnesses such as cancer.

But beyond the compelling anecdotes and hopeful ideas presented by crystal healers, there is no scientific evidence that crystal healing can treat medical conditions; in fact, a 2018 study showed that patients who used alternative therapies alone were over twice as likely to die from treatable

SCIENTIFIC SIDEBAR

Throughout this book, look for these Scientific Sidebars to define, explain, and expand upon the scientific concepts we discuss.

forms of cancer. While an extreme example, this alarming fact highlights some of the major issues with crystal-healing beliefs, namely that the alternative healing industry can exploit people's vulnerabilities by claiming to offer an affordable and effective means of treatment.

It's unlikely that you've picked up this book without any preconceptions about crystal healing. Maybe a friend told you about the crystals they used, or perhaps you saw a social media post about crystals. However this concept of energy-based healing was first presented to you, you were probably left with many questions. But it can be daunting to try to separate the seemingly plausible theories from the science fiction. In this book, we'll talk about what crystals really are and why they may sometimes appear to work for healing, why people believe they are mystical in the first place, and the hidden costs of mining and selling crystals for the crystal-healing industry.

AN INTRODUCTION TO ROCKS AND MINERALS

When talking about crystals, whether in a scientific setting or when discussing crystal healing, it helps to have a firm understanding of what crystals are, what they're made of, and how they differ from other natural materials such as rocks. Here we'll provide a brief introduction to the inner workings of our planet and how crystals come to be, and how everything from their shape to their color is determined by the geometric arrangement of particles too small to see. With these basics of geology and mineralogy, it will be clear that crystals contain no magic, no matter how enigmatic they may seem.

"If there is any 'magic' to crystals, it is the same kind found in a sunset or in the aurora borealis: a completely natural and understood phenomenon that happens to have a beautiful result."

Crystal "healers" do an excellent job of presenting crystals as mysterious objects with almost magical properties, but if your primary exposure to crystals was through crystal-healing rhetoric, what you may not realize is that there is, in fact, no mystery to crystals. We know exactly what they are and how they develop their peculiar, intriguing shapes. And because they've been researched and understood for decades, we also know that they do not emit any curative energy. But the crystal-healing industry survives on a misunderstanding of this fact; it relies on misinformation and the public's general lack of knowledge of the true makeup and properties of crystals, and takes advantage of the well-meaning enthusiasm many people have for the beauty of crystals. Pointing to their wild shapes and stunning clarity, "magic" starts to seem like a valid explanation to some.

But why does this belief exist? A good part of the reason is that geology, mineralogy, and crystallography are inherently complicated subjects. Understanding them and their terminology requires real research, and their core ideas are difficult to quickly convey. This can be frustrating for someone eager to begin learning about crystals, and that person may instead look for faster, easier answers. Crystal healers often provide such an answer: false claims that are easier and more fulfilling because they are invented in a way as to leave no questions unanswered. But the complexity of these subjects doesn't make them any less valid, nor does it mean that crystal healers get to ignore physics and rewrite the facts. Crystals have testable, measurable properties, none of which includes divine sources of power. In this section, we'll look at what crystals actually are, down to their tiniest components, to show you that you will find no mystery at their core—just an incredible story of natural history.

WHAT ARE CRYSTALS AND MINERALS?

A trick performed by a talented magician is always wondrous and mysterious until it is explained to you exactly how it was performed. With the veneer of magic stripped away, you can see how sleight of hand and clever misdirection led you to believe that something more incredible was happening. Crystals, with their enigmatic geometric shapes, rainbow of colors, and ancient origins, can appear

These astrophyllite crystals are still embedded in their host rock.

mysterious, even alien, and they remain a source of almost spiritual wonder until they've been properly explained. And, like analyzing each step of a magic trick, when we study crystals and learn more about them, we find no magic but something more interesting: a perfectly geometric arrangement of atoms.

Atoms are the *smallest* whole particles we currently understand. Everything is composed of atoms; when we break something down and analyze what it's made of—whether it's a crystal, our bodies, or entire stars—the smallest whole pieces we end up with are atoms. Atoms have innate electrical charges and consist of two primary parts: a **nucleus**, which contains **protons** (positively charged particles) and **neutrons** (particles with a neutral electric charge), and a "cloud" of **electrons** (negatively charged particles) that surrounds and orbits the nucleus.

A simplified scientific diagram of a silicon atom

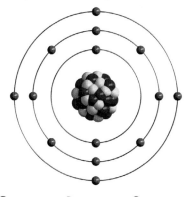

● **14 Protons**   ● **14 Neutrons**   ● **14 Electrons**

The number of protons in an atom's nucleus determines how many electrons it can hold in its orbit. The more protons and neutrons an atom has, the "heavier," or larger it is, and the more electrons it holds. As a result, atoms of different sizes have completely different properties; we call these differently sized atoms **elements**, of which there are currently 118 known (though only 94 occur naturally). You're already familiar with many elements; elements are the "building blocks" of the universe. Hydrogen is an important example as it is the lightest element and most abundant substance in the universe, and it contains just a single electron per atom. But other elements, like iron, copper, and oxygen, are likely more familiar to you in everyday life.

Much like the positive and negative ends of a magnet, atoms of various sizes are also attracted to each other. The number of electrons surrounding an atom's nucleus determines how "magnetic" it is and to what kinds of atoms it is attracted. Atoms "prefer" to have a certain number of electrons, and many can achieve their preferred number by sharing with another atom. These are called **atomic bonds**. When two or more differently sized atoms—which is to say, two or more different elements—bond together and share electrons, it is called a **molecule**. Many identical molecules bonded together in one mass is called a chemical compound.

For example, the element sodium contains atoms of relatively light weight, with 11 electrons, and the element chlorine is a little heavier, with 17 electrons. Simply put, due to their atomic structure, sodium would "prefer" to have 10 electrons and chlorine would ideally have 18. These "preferences" make these two elements reactive. When in each other's presence, sodium and chlorine atoms react by sharing the "extra" electron between them. They bond together, forming

Halite, the same mineral as table salt, consists of sodium and chlorine.

a chemical compound called sodium chloride. When sodium chloride is found in nature, we call it halite, which is a mineral that forms cubic crystals; you know it better as common table salt.

Minerals are inorganic chemical compounds formed when atoms of various elements bond together and **crystallize**, or harden. **Crystals** are the solidified forms taken on by minerals. Several factors determine which mineral will result when a compound crystallizes: the elements that are bonding, the proportions at which they combine, and the environmental conditions during crystallization. For example, the same chemical compound can crystallize in one shape under low pressure and temperatures but take on a completely different shape under high pressures and temperatures, with each result being a distinct, separate mineral species.

But how do crystals develop such perfect shapes with exact angles? That's because the way elements bond together at the molecular level are not random. When a chemical compound crystallizes, or solidifies, the molecules within it bond together at specific points determined by their internal atomic structure—atoms are compelled to bond with other atoms that can help them achieve their "preferred" number of electrons.

Much like how stacking bricks can only be done one way—one on top of the other—molecules of a certain type can only bond one way in the particular conditions present at the time of crystallization. As they bond together and expand outward in all directions, they develop a distinct crystal structure unique to each mineral species.

SCIENTIFIC SIDEBAR

A few minerals, such as gold and diamond, consist solely of a single element. In the case of gold, the element is, of course, gold, and with diamond, the element is carbon. These are called **native elements**; gold and carbon, among a few others, are particularly stable elements that don't need to bond with other elements in order to crystallize. Despite this unique trait, they are still considered minerals due to the fact that they still crystallize like any other mineral.

Crystal structures are the highly organized, symmetrical arrangements of molecules that make up crystals. The crystal structure repeats constantly in all directions, incorporating all the available molecules into a neat latticework of atoms all bonded at specific angles. Though the lattice itself is too small to see with the naked eye, the visible results of minerals' internal crystal structures are the geometric shapes we know as crystals. And it's the crystal structure of a mineral that determines all of its physical properties, including color.

This crystal is comprised of pure copper; it is a native element.

HOW, WHERE, AND WHEN DID MINERALS FORM?

With a few exceptions, minerals form within the Earth, whether just a few hundred yards below our feet or 200 miles down in the hotter parts of our planet. But because the vast majority of minerals and their crystals develop in places where we cannot possibly observe them, their formation can still seem mystical. Many crystal-healing proponents like to give vague, hyperbolic statements that crystals were "born at the center of the earth" and are "ancient beyond our knowledge" in order to help bolster their claims that crystals have healing energy, but such claims are far from accurate nor are they helpful for understanding how minerals form.

Minerals can form in a number of different ways, and often under conditions and in a scale most of us have a hard time imagining: the materials are often subjected to millions of pounds of pressure, thousands of degrees of heat, and it can take tens of thousands of years for such conditions to force atoms to bond in ways they might not otherwise. Some minerals develop only in the presence of water, some only in the absence of water, some only below a certain temperature, and some only above a certain acidity. There is a seemingly endless number of variables that affect mineral development, which makes mineralogy a diverse and fascinating area of study. While the various processes that produce minerals can (and do) fill volumes, the core principles behind the most common processes are not difficult to understand.

To understand minerals, it's helpful to know a little about the Earth and its geological structure. All known life on Earth lives on, or within, Earth's

SCIENTIFIC SIDEBAR

Put most simply, a mineral is solid mass composed of an element or combination of elements that has a distinct set of properties determined largely by the shape of its crystal structure. Crystal structures are themselves determined by how atoms have bonded together and the external forces, such as heat and pressure, acting upon them as they did.

The outer crust of earth is composed of hard, rigid layers of rock.

crust. The **crust** is the outermost layer of our planet; it is rigid and stiff, composed largely of hardened, cooled rock. All of our continents, mountains, lakes, and oceans are part of, or sit upon the crust. The crust is also the thinnest layer of the earth, averaging only about 18 miles at its thickest and just 3 miles at its thinnest. By comparison, the next layer down, the mantle and its constituent layers, is close to 2,000 miles thick, making up 82 percent of the planet's total volume.

The **mantle** has sub-layers, the most prominent being the asthenosphere and the lower mantle. The asthenosphere is hot and under such pressure that rocks there are semi-solid and malleable—these partially melted rocks slowly deform, stretch, fold, and compress over millions of years. The lower mantle, on the other hand, is even hotter, but it is under such immense pressure that the rock there can't deform like those in the asthenosphere, and so it is more solid.

At the center of the earth is an outer **core** that surrounds the inner core. The outer core consists almost entirely of superheated iron and nickel, swirling around the inner core in a fluid state. The inner core is also composed primarily of iron and nickel, but under such pressure that it remains solid. The heat of the core drives the rock-forming forces of the mantle above it.

The crust is divided up into huge sections called **tectonic plates**, which "float" and move upon the soft, hot asthenosphere. The movement is generated by convection within the mantle; just as happens in water or air, the hotter rock rises while the cooler rock sinks. The plates on the surface are thus in constant, but very slow, motion, sliding past each other and pushing into each other. Day-to-day, this action creates earthquakes and volcanic eruptions, but over thousands of years, continents can move around the globe, and some plates are pushed upward to become mountain ranges while others are driven downward and disappear into the Earth. As those plates descend deeper and deeper to where it is hotter, they eventually begin to soften and melt, becoming part of the mantle and further driving the earth's geological changes.

THE INNER LAYERS OF THE EARTH

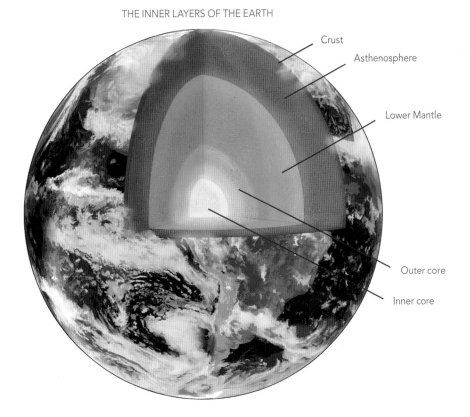

Crust

Asthenosphere

Lower Mantle

Outer core

Inner core

SOME WAYS MINERALS FORM

The great pressures in the mantle can push large bodies of molten rock called magma upward into the crust. This hot material contains a mixture of minerals dissolved within it, a jumbled mass of molecules too hot and energetic to bond together. But eventually the magma begins to cool and the minerals within it can begin to crystallize. Different minerals have different melting points, so as the mass cools, certain minerals will crystallize first and the later ones have to fit in around them. This is called **magmatic crystallization**, since it begins with magma, and it is one of the primary ways in which minerals are formed within the earth.

As magma is pushed upward into the crust, it can contact groundwater, heating it and enriching it with minerals. As this superheated water and steam rises up through the older rocks above the magma, it deposits the dissolved minerals within cavities and openings in the rock. The longer the rock remains inundated with the mineral-rich water, the more mineral molecules are introduced into the cavity where they can concentrate and bond together to form larger crystals. This is called **hydrothermal activity** and is another prominent way that minerals form, especially those rich with metals like copper and gold.

This example of porphyritic basalt shows how some minerals, namely these feldspars, crystallized before the rest of the rock.

Some minerals form via **precipitation** of dissolved minerals from water, even on the Earth's surface. As mineral-rich water dries up, the minerals within it are left behind, and if the concentration is high enough, well-formed crystals can develop. On the shores of the Dead Sea and other super-salty bodies of water, for example, perfect halite crystals can form as the water evaporates, exactly as they do if you let a glass of saltwater dry up at home. Many minerals that form this way can do so very rapidly, sometimes in a matter of hours.

Not all minerals form "fresh." Some form when preexisting minerals are changed and altered due to external forces. As some tectonic plates are forced downward into the heat and pressure of the mantle, the minerals in them partially melt and soften, which enables them to change completely. As molecules from these minerals are compressed, they may recrystallize in an entirely different order, and may incorporate other elements in the process. This is called **recrystallization**, and it is how many gemstones, such as garnets, are formed.

In most of these processes, and in the many others that also occur, a few rules of thumb generally apply. In most cases, the longer a mineral has to form—which is to say, the longer the conditions for its growth remain stable and conducive—the bigger its crystals can become. For minerals that formed in cooling magma, for example, those that crystallized very quickly near Earth's surface may

This gemmy andradite garnet formed within fibrous chrysotile asbestos in a metamorphic rock environment.

remain microscopic in size, while others that cooled very slowly, deeper down, can grow to inches or even feet in size.

The mineral-forming processes presented here are just a handful of the many that take place within the earth, and their descriptions have been greatly simplified.

WHEN MINERALS FORMED

But the question remains: when did minerals and their crystals form? For some mineral specimens, the answer is millions or billions of years ago, as metaphysical practitioners love to acknowledge. The agates of the Lake Superior region, for example, formed 1.1 billion years ago, before life existed on land. But some agates found elsewhere aren't even 100 million years old. Other minerals, such as halite or gypsum, can literally begin to crystallize overnight. Many spectacular gypsum crystal specimens on the market were formed in mines after the construction of mining tunnels; some incredible examples exhibit gypsum crystals coating miner's tools that were left in standing water within the tunnels! This fact is often left unexplored by many crystal-healing authors in favor of the blanket-description of all minerals being "ancient." Some crystals aren't; some are brand new.

❝Crystals have measurable properties. They aren't divine sources of power.❞

WHAT ARE ROCKS?

Rocks are mixtures of minerals that have crystallized or otherwise joined together in a large mass. Some rocks consist primarily of a single mineral while others contain dozens. There are three primary groups of rocks, which are classified by how the rocks form: igneous, sedimentary, and metamorphic.

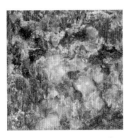

Granite, an igneous rock

Igneous rocks are those that form directly from the cooling and solidification of magma (molten rock within the earth) or lava (molten rock that has reached the earth's surface). This can happen over millions of years deep within the earth to form coarse-grained rocks like granite, or it can happen very quickly on the earth's surface when a volcano erupts lava into the cool atmosphere, forming fine-grained rocks like rhyolite. Whether you can see the individual minerals with your naked eye or not, these rocks consist of interlocked mineral crystals and grains, formed during magmatic crystallization, as described above. Igneous rocks are a primary environment for fine crystals.

Sandstone, a sedimentary rock

Sedimentary rocks form when particles of other rocks, minerals, or organic matter, such as sea shells, accumulate and compact together to form a solid mass. This typically occurs in bodies of water where small worn-down pieces of rocks and minerals, particularly quartz, settle and are locked together by pressure and/or by dissolved minerals crystallizing between the grains and "gluing" them together. Sedimentary rocks can produce finely crystallized minerals, but fewer varieties than are found in igneous rocks.

Gneiss (pronounced "nice"), a metamorphic rock

Metamorphic rocks form when preexisting igneous, sedimentary, or other metamorphic rocks are subjected to heat and/or pressure within the earth, compressing or softening them enough for their constituent minerals to change, reorganize, and recrystallize. Metamorphic rocks often develop a layered or flaky texture, proof of the crushing forces they've endured. They are also a primary environment for crystals, particularly gemstones like garnets and rubies.

To sum up, rocks contain minerals—and can contain large, well-formed crystals—but rocks themselves are not minerals and they are certainly not crystals.

WHY DO CRYSTALS SEEM SO RARE?

Anywhere you go, minerals are somewhere underfoot. As constituents of rocks, minerals make up the entirety of the earth's crust and can be found everywhere, if you take the time to look. But if you're wondering why you don't find museum-worthy clusters of perfect crystal points everywhere you look, that's a different question entirely.

Minerals and their crystals will continue to form and increase in size as long as all the conditions for their growth remain constant—a source of molecular material, constant pressure and temperature. But the amount of space available to the growing crystals is important, too. In a large cavity, crystals may have plenty of room to grow to large sizes and have perfect shapes. But in a small cavity, the growing crystals may rapidly fill the entire opening and develop a tightly packed cluster with no visible crystal points. Only in larger cavities will the most stunning crystals form, and unfortunately, small cavities are far more common than large ones. This means that while minerals and their crystals are found nearly anywhere, they are most often tightly packed into the structure of rocks, often barely noticeable.

FACT VS. MYTH

Perhaps now, after reading about the elements, chemical compounds, and minerals that all come together to form rocks, you can see why many crystal-healing authors choose to skip over these facts. It is, admittedly, a complicated subject that requires quite a bit of prior knowledge to grasp, and many authors don't bother to delve into it at all, perhaps because they don't understand it themselves or because it is much easier to make claims of mystical origins and leave it at that.

The ample space in this geode allowed this calcite crystal to grow to a fairly large size. Specimen courtesy of Chris Cordes

And that is the primary difference between crystal healing authors and actual scientists—crystal healers are able to tell a neat story that isn't concerned with getting the facts right. For some, this may be more valuable and satisfying. Such tales bolster the mythology and hyperbole surrounding crystals, but the stories they tell also demonstrate their lack of understanding of crystals. By coming up with their own unbelievable, unreasonable stories of the "power of nature," they unwittingly overshadow the incredible natural forces actually at work.

If there is any "magic" to crystals, it is the same kind found in a sunset or in the aurora borealis: a completely natural and understood phenomenon that happens to have a beautiful result. When we look at crystals, we don't need to write stories about their origins; they already have their own.

SCIENTIFIC SIDEBAR

Some definitions to remember:

Mineral Inorganic chemical compound that has a distinct crystal shape

Crystal The hard geometric shape taken on by a solidified mineral defined by its chemical composition

Rock A hard mass consisting of a mixture of minerals

Epidote is popular for its often boldly shaped crystals and its rich color.

Phenakite is a rare mineral strongly associated with crystal healing. Many metaphysicists insist it is among the most powerful of minerals.

Pyrite, also called "fool's gold," is one of the most abundant metallic minerals.

Enigmatic and interesting crystals, like this specimen of aegirine, have captured imaginations for millennia, but are far from being magical.

SCIENTIFIC SPOTLIGHT: ATOMIC BONDS

When atoms bond, they do so in certain positions based upon how they relate to the atoms of the other elements in the bond. When discussing minerals and their crystal structures, these positions are often visualized as simple geometric shapes. Quartz is a good example; it consists of four oxygen atoms bonded to one silicon atom,

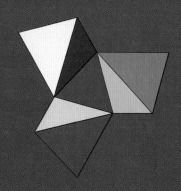

Three silica tetrahedra bonded in their preferred orientation

a compound also known as silica. This arrangement of the four smaller oxygen atoms around the single larger silicon atom takes a four-pointed shape called a tetrahedron, which is illustrated as a triangular pyramid.

As silica molecules (in the form of tetrahedra) bond together, they attach to each other at their corners, each rotated slightly in relation to the next. They don't bond together randomly, as their innate charges cause them to be attracted to each other in an orderly manner. And they ideally bond

Countless silica tetrahedra are bonded within every quartz crystal.

at more or less equal rates on all sides of a growing crystal. The resulting lattice is complex but distinctly organized; the overall shape is distinctly hexagonal, or six-sided. This internal, microscopic lattice determines everything about a crystal, including its shape.

You may be looking at this quartz crystal lattice diagram and wonder, "How does that jumble of molecules make a perfect, six-sided crystal?" The answer becomes a little clearer as the perspective changes. A close-up of some silica molecules shows how their orientation changes from one to the next when bonded, but the overall pattern they create as a whole cannot be appreciated until we pull back.

Viewed from top-down, we can see the near perfect hexagonal shape of this quartz crystal.

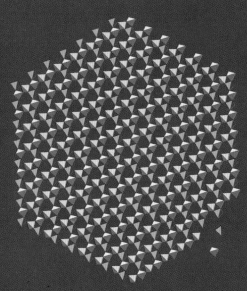

Silica tetrahedra bond at specific positions that produce a hexagonal lattice, with additional molecules taking their place in the growing structure.

Because silica tetrahedra in the surrounding material (often hydrother-mal water) are attracted to the developing crystal from all directions, they assume their positions in the growing lattice and begin to develop the characteristic quartz shape. This is similar to taking a small snowball and rolling it around in the snow; it will accumulate more and more snow, growing in size while retaining its spherical shape. Similarly, the adher-ing silica tetrahedra build upon themselves and contribute to the already established quartz structure. When we "zoom out" and view the lattice from above, for example, we can start to see the orderly hexagonal shape that we know from well-developed quartz crystals.

How do we know that this is the shape that the molecules take as they bond? Because we can view it directly through a process called transmis-sion electron microscopy (or TEM), which allows us to see how atoms bond and build upon each other.

Quartz crystals often grow in clusters.

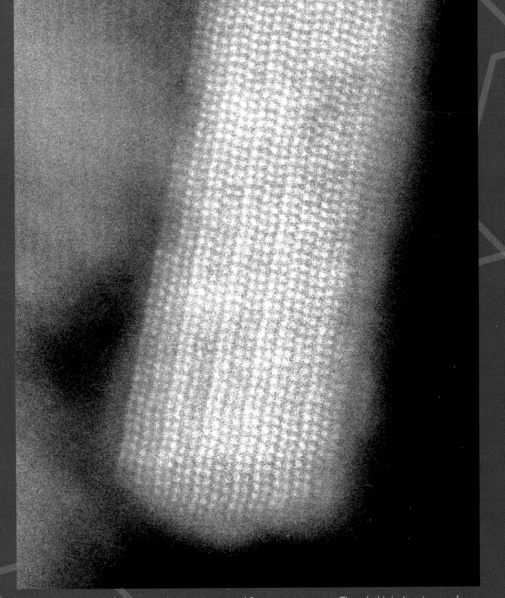

This is a TEM image of a tiny crystal measuring around 5 nanometers across. Though this isn't an image of a quartz crystal, you can still see how a TEM image reveals the highly organized arrangement of atoms (the white dots) in a mineral.

AN INTRODUCTION TO METAPHYSICS

Crystal healing falls under the umbrella of New Age Metaphysics, the philosophy that all things in the universe are connected by unseen energy. Through "open-mindedness" and "attuning ourselves to nature," it is said that we are able to use this energy to do a great many things, including healing. But the modern version of this belief has roots in ancient history, long since appropriated and transformed into the crystal-healing practices we see today. In this chapter, we'll look at what crystal healing is, where it comes from, and who believes in it.

"You don't have to be open-minded for the earth to revolve around the sun—it's true regardless of your personal beliefs—but you do have to be 'open-minded' to believe that holding rose quartz will improve your love life."

WHAT ARE NEW AGE METAPHYSICS?

Anyone who believes and practices crystal healing is, by extension, a believer in New Age metaphysics. Even if you're unfamiliar with what that means, you've no doubt heard the term used, usually within the context of unexplainable phenomena and mysticism. But New Age **metaphysical philosophy** is a system of beliefs built around the idea that all things in the universe are connected via unseen energy. By understanding and becoming in tune with this interconnectedness, it is said that we can harness

Crystal healers claim that reverence for crystals can attune our bodies to the universe's energies.

the energy surrounding us to understand the nature of reality and to do incredible things—things like communicating with the dead, using the positions of the stars to predict the future, and, of course, using crystals to heal illness, pains, and emotional troubles. These are all metaphysical practices that share a core concept: because everything is interconnected within a vast network of energy, we can learn to transcend our human abilities and both sense and tap into this energy. Crystals, for example, are usually claimed to be a physical manifestation of Earth's energy and can be used as a sort of "conduit" for earthly healing power.

But metaphysics haven't always been tied to mysticism. Historically—going back at least to Aristotle in the 4th century B.C.—metaphysics is a branch of philosophy concerned with exploring the very structure of reality and experience. The concept of how our minds relate to and interact with matter has long been a core issue. But metaphysics have never been a science, as most of its principles are not testable or verifiable—though that has not stopped people from appropriating its complex ideas and turning them into a set of beliefs. The abstract ideas and unanswerable questions posed by famous philosophers were later appropriated by New Age thinkers. These concepts, coupled with lots of modern scientific jargon stripped of its context (e.g., quantum, matrix), are the basis of some of the more fantastical viewpoints of our universe that we encounter today. By disguising questionable concepts like crystal healing and astrology under the imprimatur of "metaphysics," proponents of these practices have assumed a position of legitimacy.

Today, there are many different flavors of modern New Age metaphysical beliefs. For many practitioners, this is part of the appeal: they are not concerned with preexisting religious dogmas, proven scientific findings, or manmade laws, but rather are more interested in their own interpretations of reality. Particular emphasis is placed on finding a physical or spiritual "balance"—whatever that may mean to you—with the universe. And all of this can be achieved by remaining "open minded" and having faith in the universe.

And therein lies the crux of New Age metaphysics: the concept of open-mindedness. You don't have to be open-minded for the earth to revolve around the sun—it's true regardless of your personal beliefs—but you do have to be "open minded" to believe that holding rose quartz will improve your love life. In metaphysics, "open-mindedness" is treated as both a justification for unbelievable claims and as a counterargument to any criticism of crystal healing. The concept of open-mindedness in this case is akin to blind faith.

Crystal healing is a prime example of how New Age metaphysics works— its proponents claim that by simply keeping an "open mind" and believing in the earth's "healing energy," "vibrations," or other buzzwords, you can use crystals to ease pain and anxiety and enjoy a healthier lifestyle. Thus modern metaphysical concepts are inherently abstract and subjective; what they "actually are" is whatever you want them to be.

Of course, such claims aren't true, but the fact remains that many people have found this idea to be a reality they'd rather believe. But how did this all come about?

Minerals, flowers, essential oils—some of the typical trappings of metaphysical practices

WHERE DO CRYSTAL-HEALING BELIEFS COME FROM?

The use of crystals and stones to treat ailments and improve the user's life dates back thousands of years. Modern crystal healing has roots in Indian Hinduism, Tibetan Buddhism, Indigenous American cultures, and Northern European Paganism, among many others, all of which have historically seen some use of stones in their practices and rituals. Today, elements of each of these cultures' traditions have been appropriated, reinterpreted, and blended together into the crystal-healing techniques used by both devout metaphysical practitioners and more casual alternative-medicine seekers alike.

The melding of multiple cultures' traditions in crystal healing was largely a result of the counterculture and New Age movements of the 1960s and 1970s. Along with the wider anti-war and anti-establishment movements, there were also disparate movements to reject modern medicine and increasingly modernized technology in favor of more natural and authentic ways of living and healing. Meditation, yoga, and other "ancient" practices promoted a connection to nature, inner peace, and individuality that was perceived as absent from Western culture at the time. The New Age counterculture was the result, and its gurus adopted and appropriated any culture's traditions that fit their ideals.

Many New Age followers and modern crystal healers place particular emphasis on the most ancient stone-centric traditions. But these connections are tenuous, at best; many of these beliefs are based on embellished histories or those that are outright invented. For example, in traditional Chinese medicine, crystals haven't historically played a large role. However, association of crystals with Chinese medicine has grown in recent years due to the New Age interest in combining Eastern folk healing with other alternative medicine beliefs and practices. In particular, the concept of "qi" (or vital energy) has been appropriated by crystal healing adherents. Similarly, the traditional Chinese meridian maps used in acupuncture have begun to be appropriated for use with modern crystal-healing techniques.

"Many New Age beliefs are based on embellished histories or those that are outright invented."

This image, showing the supposed chakra points, is typical of the mystical language and imagery presented by crystal healers.

In crystal-healing guides, particular attention is often placed upon ancient Egyptians, Greeks, and Romans, all of whom viewed precious stones and crystals with a sense of reverence. But modern metaphysicists often speak for these ancient cultures, making assumptions and exaggerations about how they may have used and viewed stones. The Egyptians did decorate their tombs and sarcophagi with stones and gems, believing them to facilitate transition to the afterlife, and folk healers in ancient Greece and Rome attributed healing properties to certain materials and stones, and these were sometimes passed down by historians such as Pliny the Elder. Such beliefs continued all the way through the Roman era.

These ancient writings include recipes and instructions for how to grind, heat, or soak various stones in certain combinations to cure a variety of ailments. But they often used less-than-clear terms, with numerous caveats, suspect interpretations of results, and inclusions of minerals known today to be toxic. Take the legendary Hippocrates, for example: around 400 B.C., he recommended a poultice containing (among other things) lead and arsenic as a treatment for ulcers. An ancient remedy, to be sure, but one we would never try today. Still, this doesn't stop many modern metaphysicists from quoting ancient sources to strengthen their claims.

A great number of aspects of modern crystal healing have been appropriated from the Indian Subcontinent. In India, Hinduism has long placed special emphasis on using stones to align, balance, and charge chakras, the points of power along the spine at which our physical and spiritual selves are said to intersect. Knowing when and how to use crystals at chakra points is supposed to keep them balanced, ensuring that none become too powerful or too weak; imbalanced chakras are said to cause all manner of maladies. The Vedas, which are the ancient sacred scriptures of Hinduism, and the Puranas, the later encyclopedic texts, are the sources of this belief and provide the basis for why crystals are said to have divine origins and innate powers. The ever-prominent connections made by metaphysicists between crystals and Tibetan Buddhism (often incorrectly associated with Hinduism) are largely a modern marketing invention; Tibetan Buddhism traditionally includes no explicit use of crystals or stones, but does utilize them to some degree through the use of malas, or strands of beads used in meditation. Along with beads of stone and crystal, wood, seed, bone, and metal beads are also used in malas to count recitations of mantras and prayers.

A traditional mala bracelet made of stone beads.

In North America, several indigenous tribes have had their own long-time uses for crystals and stones, the knowledge of which was traditionally only passed on through storytelling and oral teachings. Navajo, Apache, Cherokee, and other tribal nations have incorporated crystals into their healing practices and place particular emphasis on meditation and respect for the earth. Crystals are often handled with great care, and are often worn or carried to receive their purported benefits. In many cases, these traditions are carried out with more-or-less common stones, not necessarily colorful or well-formed crystals, seemingly antithetical to much of the modern crystal-healing milieu.

Several of today's crystal-healing practices also originated in Northern Europe, particularly in Scandinavian Nordic culture and early Anglo-Saxon

A modern-day example of a Pagan crystal grid

culture in England. These Pagan traditions are often closely associated with astrology, or the belief that the positions of the sun, moon, and star constellations can be a source of divine information. Certain stones associated with celestial objects, often arranged into grids or layouts around the user, are used to receive energy and make predictions, decisions, and even to speak to the dead. Many practitioners today have also incorporated use of a "wand"—whether it be a large crystal or a stone carved into a wand shape—which is said to facilitate the activation of the grid surrounding the user.

These centuries-old traditions are examples of **folk remedies**, or folk healing, the practice of which arose from a desire to understand the world and how it affects our bodies before sufficient knowledge existed to help explain it. These cultures resorted to superstitions and trial-and-error techniques when attempting to explain and treat illnesses, and their methods of healing developed from whatever successes they found. Before our current understanding of disease, for example, many illnesses were said to be the result of anything from evil spirits to angered gods to flaws in the afflicted person's character. A folk healer may have tried any number of remedies, including those involving crystals, and any that may have yielded a perceived positive effect would be further practiced and explored. But, lacking a full understanding of the processes at work, positive benefits were often misattributed. Using crystals to treat someone who has a minor illness, such as the common cold, may appear to work because our bodies are able to cure themselves of such a minor illness. If the use of crystals happened to coincide with the patient's natural immune system response to the illness, a folk healer could attribute the lessened cold symptoms to the use of crystals. Similarly, use of a crystal could also appear to cause a patient's condition to improve because of other aspects of the treatment, such as the patient utilizing deep breathing to achieve relaxation. Thus the tradition of folk remedies continued, with many "healers" using methods that only appeared to work. For more information on how this way of thinking arises, see page 102.

A RECENT HISTORY OF CRYSTAL HEALING

Ancient civilizations and eastern cultures weren't alone in incorporating folk remedies such as crystal use into their healing practices. For centuries these remedies held a role in Western cultures and European organized religions, too. But as the world began to modernize, and the medical profession and hospitals emerged and steadily improved over the centuries, folk remedies were gradually diminished in importance in favor of evidence-based medicine and research. These new practices and techniques focused on the body and its natural healing abilities rather than external magic. Significant advancements in the safety, reliability, and effectiveness of medical treatments were being made, and folk remedies like crystal healing played much less of a role.

The importance of crystal healing may have been diminished, but the practice never disappeared—it merely changed. Much of the modern crystal-healing movement has its roots in the Victorian Era and corresponding craze with the paranormal, magic, ghost stories, and the like. Largely an activity of the wealthy and the upper class, crystal gazing, crystal grids, occult rituals, and fortune telling were popular pastimes. Such activities were frequently performed behind closed doors; what were once everyday practices had become exciting taboos. These activities, merely a curiosity to some but a serious interest to others, grew in popularity

Crystals, burning herbs, and incantations are all examples of folk remedies.

and became more widespread. Soon, occult spokesmen and fortune tellers began to use pseudoscience, compelling anecdotes, and their vibrant personalities to gain followers, sell products, and sell tickets to view demonstrations. Even prominent scientists and public figures bought into the excitement of metaphysics: psychologist Carl Jung gave credence to metaphysical practices and ideas, but more influential were popular writers and self-proclaimed "prophets" like Edgar Cayce, who promoted them as a lifestyle.

"When medical treatments are deemed inadequate, people may look for other options, including crystal healing."

As the metaphysical movement continued to progress, the alternative reality provided by its mysticism caused some to take their beliefs to extremes. Stories of sunken continents and lost civilizations, sources of ancient power, and alien planets colliding with Earth are just a few of the wildest ideas purporting to be a part of metaphysics. (These ideas were also soon enmeshed in popular culture, leading to the rise of science fiction, adventure novels, and the like.)

But it was the more "down-to-earth" aspects of metaphysics, such as crystal healing, that found the most footing within modern culture.

WHY PEOPLE TURN TO CRYSTAL HEALING

The Victorians may have re-popularized many aspects of metaphysics, but the modern resurgence of metaphysical ideals, especially crystal healing, owes much more to general public concern over perceived limitations and problems with modern medicine.

One of the primary reasons folk healing has held such a long-lasting appeal is that until relatively recently medicine simply couldn't offer effective treatments or cures for many common diseases and maladies. Plague and malaria, tuberculosis, childhood diseases like measles and pertussis, and even something as simple as an infected wound were all potentially life-threatening issues for which modernized medicine was slow to cure. The situation finally began to improve more rapidly in the late 19th and early 20th centuries with the acceptance of germ theory, which states that microscopic organisms entering the body can cause disease. This knowledge led to a broader understanding of how to prevent disease. Numerous ground-breaking treatments, including vaccines and antibiotics, soon followed. Combined with improved nutrition and sanitation, this movement into modern medicine led to people living longer, healthier lives. As the world continued to claw its way into the modern scientific era, fewer and fewer people placed their faith in folk remedies.

But scientific and medical progress can be uneven. While great progress— even outright cures—has been made on many diseases, others, such as

Crystal-healing beliefs and activities have begun to grow in popularity for a wide variety of reasons, including a perceived increase in connection to nature.

some types of cancer, have a long way to go. These more challenging diseases may have seen little to no medical advancement despite modernized scientific research. For those afflicted with these diseases, modern medicine can become a source of frustration instead of hope.

The experience of modern medical care is another issue. While the healthcare provided in the modern hospital setting is better, safer, and more controlled than ever before, the experience can be impersonal, intimidating, and overwhelming. Coupled with the increasing cost of medical care, the complexities of insurance providers, and the sometimes-impersonal bedside manner of doctors, would-be patients may find themselves resenting and avoiding hospitals altogether.

When there aren't adequate treatments, or the existing ones are inaccessible, expensive, or have pronounced side effects (such as those with chemotherapy), people may look for other options. They may feel that modern medicine has failed them. And this feeling can—and does—make patients think their only option is exhausted, which can cause genuine despair and anger that fosters a general distrust in science and medicine. This can in turn result in other negative lines of thought, such as susceptibility to conspiracy theories claiming that modern medicine or "Big

Pharma" is holding back cures for financial gain. These ideas can lead frustrated patients to reject medicine entirely in favor of something that can seemingly provide the answers they seek. Crystal healing is one remedy that people have turned to as a result.

Crystal grids—geometric layouts of stones said to facilitate healing—are a symbol of New Age styles of thinking.

METAPHYSICS AND CRYSTAL-HEALING BELIEFS TODAY

While the New Age movement was initially viewed as a fringe subculture, it grew in popularity until the 1990s, when it began to fade somewhat in favor. But its ideals didn't disappear. The rise of the internet rapidly changed how alternative medicine and metaphysical beliefs were disseminated and shared, continually evolving and being adapted to individuals' tastes. Today, crystal healing is more popular than perhaps ever before, free of any stigma the "New Age" label may have once provided it. Now, not only is crystal healing increasingly widespread, but many adherents of the modern movement are well-educated members of the middle and upper classes.

There are several reasons why crystal healing is popular today. First, alternative medicine provides a patient a semblance of control and power when in an otherwise powerless medical position. Some forms of chronic pain, for example, may not respond to modern medicines, or may go undiagnosed, but an affected individual may find some form of comfort in attempting to treat themselves with crystals. No actual change in the patient's condition is taking place, but the sense of agency in treating themselves may alleviate some distress.

Secondly, the internet has revolutionized how information is created and disseminated, and all of that information can be taken at face value, if the reader so chooses. This has been called the "privilege of subjectivity." When so much unverified information from countless sources is made available, it is up to the reader to decide what should and shouldn't be believed. When each source makes claims about what is true, the reader must show a willingness to challenge the veracity of the content. But when faced with a deluge of content that is misleading, or outright false, readers are increasingly opting to "decide for themselves" and go along with ideas that align with their previous beliefs. This is known as confirmation bias, and will be discussed in detail on page 60.

"The modern use of crystals for healing may have originally arisen from a rejection of institutionalized medicine, but today it is a symptom of the normalization of mysticism and pseudoscience via untrustworthy sources on the internet."

This privilege of subjectivity is also a privilege in the traditional sense, as most adherents to crystal healing can afford to change their beliefs (both socially and financially) to seek out whatever truth they wish. The entitlement to a reality of one's own choosing is a symptom of this privilege and a way of seeking individuality and power in one's life.

Finally, and perhaps most importantly, the prevalence of crystal healing today, particularly on the internet, is due to the fact that alternative healing has been normalized. No longer relegated to New Agers or rural folk healers, crystal healing has found its way into the homes of average people of all ages and beliefs. Due to the prevalence of positive information about crystal healing online (with little to no active criticism of crystal healing found online), many people no longer compartmentalize crystal healing as an odd, fringe activity and instead view it as a viable option for medical care. In many cases, it is being used as a supplement to modern medicine, but as it has become more popular, a smaller percentage of users are using crystals and other untested practices as their only forms of medicine. But when polled, average practitioners of crystal healing today make little to no claim of disillusionment or frustration with modern medicine as the New Agers once did. Instead, many state they began using crystals because they saw crystal healing simply as another option available to try. The modern use of crystals for healing may have originated from a rejection of institutionalized medicine, but today it is a symptom of the normalization of mysticism and pseudoscience via untrustworthy sources on the internet.

This brings us to the odd paradox of the twenty-first century: we have the most educated populace in history, and the most advanced science and medicine the world has ever seen, yet more and more people are turning to unproven methods of folk healing. We disseminate scientific research more quickly than ever before, yet the wider populace has no better understanding of how science works.

For all our incredible scientific progress, an intense specialization has resulted, which can make those outside the scientific fields feel excluded. Consequently, laypeople often describe science as complicated or even as sounding like "a different language." The extent of the specialization of modern science can make approaching a subject—even a simple one— daunting and difficult, at least at first.

Clearly, communication about science has lagged behind our actual scientific progress. Without proper venues and experts willing to help the wider populace to understand complex concepts, the average person is often left to their own devices, literally: Facebook, YouTube, Twitter, and Instagram are all widely used as sources of information both real and invented.

Calcite—a common and often beautiful mineral

This can help explain the rise of conspiracy theories and the corresponding turn toward metaphysics. Belief in New Age metaphysics is certainly convenient; if you look hard enough, its "findings" will neatly meet all the needs a user may have.

And since there are far fewer sources that challenge and refute such pseudoscience writers' claims with rational thought and facts, a curious but less-informed reader may be more inclined to believe them. The very fact that crystal healing has been normalized in so many people's minds is direct evidence of this. But the importance of evidence-based thought has never been greater; a number of adherents of alternative medicine, including crystal healing, have sadly died from illnesses that were treatable by modern medicine.

Practitioners of metaphysics make a tremendous number of claims. Sometimes their claims are inaccurate from the start; crystal healers often attribute the stunning geometry of crystals to inexplicable earthly magic, for example. Other claims are deeply rooted in the alternative medicine philosophy that ways of life that are seemingly more rustic, holistic, organic, and even ancient, have more value than modern scientific discoveries. All of these claims attempt to bolster the idea that folk traditions are more "real" and "wholesome" than the advancements made in the past several hundred years of scientific research. The message is that folk remedies, such as crystal healing, should be championed as better cures.

But any claim—especially ones as grandiose as the ones made by crystal healers—cannot be taken as true until rigorously tested and supported. To paraphrase renowned science communicator Carl Sagan, extraordinary claims require extraordinary evidence; those without such evidence are pseudoscience and should be approached with a deep skepticism.

Smoky quartz is often cited as a powerful healing mineral.

SCIENCE AND THE SCIENTIFIC METHOD

One of the foremost differences between a scientific claim and an unscientific one is the possibility of showing that the scientific claim is wrong. The unscientific claims of New Age metaphysics, including those in crystal healing, are problematic because they present arguments that can't be observed, tested, or refuted, and therefore can't be trusted. Good science, on the other hand, is performed rigorously, exploring a topic from all angles in measurable, repeatable terms, including those that may even disconfirm a previous belief. This chapter will teach you the basic principles of good science and how they contrast greatly with those of metaphysics.

WHAT IS SCIENCE?

The word science translates literally to "having knowledge." There are many different branches of science, ranging from the physical sciences (e.g., astronomy, chemistry, physics, geology) to life sciences (e.g., biology, medicine, psychology, neuroscience) to social sciences (e.g., sociology, economics). The aim of science is to identify principles, or a set of conclusions, that help to explain the world around us and make predictions about the future.

Contrary to how science is depicted in the media, scientific breakthroughs rarely happen overnight with some dramatic new insight. Instead, science is a slow, daily grind toward more-accurate knowledge. It requires decades of hard work involving rigorous observation and experimentation, which is all part of what is known as the scientific method. Science does not aim to discover irrefutable facts, but instead aims to put forward evidence and conclusions that continue to be scrutinized, and improved, through repeated testing.

When we are seeking information to make decisions, we should try to determine the accuracy of the information provided. Statements that are scientific should carry more weight in our decision-making process than those that are not. Returning to Statements A and B, both provide information about the properties of crystals. Does one seem to be more scientific than the other? In order to be categorized as a scientific claim, it must satisfy three basic criteria: (1) it must be specific, (2) it must be measurable, and (3) it must be falsifiable. Statement A meets these three criteria. This statement is specific because it offers a clear definition of the composition of crystals. Second, the statement is measurable because we have the tools necessary to determine which molecules make up the crystal structure. Last, the statement is falsifiable because there is a clear set of conditions that must be met in order for the statement to be supported. We must be able to observe molecules, and these molecules must be bonded together in a certain way. Statement A therefore is scientific because it meets the criteria of specificity, measurability, and falsifiability.

WHAT IS PSEUDOSCIENCE?

New Age metaphysics and its associated practices and beliefs, including astrology, homeopathy, and of course, crystal healing, are pseudosciences. This means that while they make claims of being scientific, followers of New Age metaphysics make claims that can't stand up to rigorous testing. Statement B, for example, falls within the realm of pseudoscience because its definition of crystals fails to meet the scientific standards of specificity, measurability, and falsifiability.

What really differentiates scientific claims from those of pseudoscience is falsifiability. Scientific claims are exactly that—claims—and they can then be confirmed or refuted by repeated testing. But many of the claims of crystal-healing adherents are simply untestable. For example, amethyst is said to help its user attain their goals. In cases such as this, the claims themselves are subjective and unspecific, and there's little or no specificity in describing the mechanism behind how amethyst might accomplish such a task. How, exactly, would amethyst help its holder attain personal goals? Since this is not an observable, measurable property of amethyst, we find ourselves in the absence of a way to confirm or deny this crystal-healing claim. However, crystal healers would tell you that this claim is true merely because there is no easy way to show that it is false. Because of this habit, we find ourselves awash in personal reports, such as "I tried this crystal, and it worked!" But anecdotal evidence and personal experiences tend to produce belief, not science.

THE SCIENTIFIC METHOD

Science is a continuous process of forming new ideas and testing how well they explain our reality. Without realizing it, we might approach everyday problems in a scientific way. One day, you might think of a new commuting route to work. Your next step might be to test it out and compare it to the old route. To know for certain that this new route is better than your old one, you would want to test it multiple times under different conditions. For example, you may discover that if you leave your house too late, you will get stuck behind a school bus. After gathering enough evidence and testing your idea under different circumstances, you may then decide to take the new route to work every day.

Testing ideas by gathering evidence in a systematic manner is also known as the scientific method. Though science in practice has been much messier, the basic elements of this rigorous and cyclical process can be summarized in a few steps.

Three questions to ask yourself to tell if a claim or belief is grounded in science:

1. Is it Specific? Does it describe a particular object, relationship, or event in concrete terms?

2. Is it Measurable? What are the appropriate tools/instruments available for testing the validity of the statement?

3. Is it Falsifiable? Does the claim identify the conditions that must be met for it to be supported?

THE SCIENTIFIC METHOD: A CYCLE

1
IDENTIFY
Locate a problem or observation in need of an explanation

2
OBSERVE
Gathering information about the problem

3
GENERATE HYPOTHESES
Formulating explanations (hypotheses) regarding the problem

4
TEST HYPOTHESES
Conducting tests or experiments and gathering data to see which, if any, of the hypotheses provide a resolution to the problem

5
CONCLUDE
Determining whether or not your hypothesis was supported and recording findings for further explorations of the problem

6
REPEAT
Identifying new problems or making new observations to more thoroughly test the accuracy of the conclusion

MEET LOUIS PASTEUR

To understand how the scientific method has been applied in real life, let's take a look at a famous example from the history of science and medicine: Louis Pasteur (1822-1895) and the introduction of the rabies vaccine. A lifelong scientist, Pasteur made several significant scientific breakthroughs, particularly in the field of microbiology. He identified that the cause of spoilage in milk, beer, and other foodstuffs was the result of microorganisms and developed a method to prolong the shelf life of foods—a process that became known as pasteurization. After twenty years of careful experimentation, research, and hypotheses, he turned his attention to microorganisms that affected people, namely anthrax and rabies.

Louis Pasteur

Rabies is a fatal viral disease for which there was no cure at the time. After building upon his own research and observations from his earlier work, combined with the discoveries of his contemporaries, Pasteur hypothesized that infectious diseases could be cured by exhausting the resources an infective agent (such as a virus or bacteria) needed to survive. As it happens, he was wrong; we know today that immunity from a disease comes after our bodies' immune systems are exposed to an infective agent and "learn" how to combat it.

But Pasteur's hypothesis still led to progress. Through a long series of tests and experiments, he isolated samples of rabies from infected dogs' brains and sought to weaken the infective agent (which we now know to be a virus) by using the material to infect rabbits. The infected rabbits' spinal columns were then exposed to air. Pasteur's theory was simple: if the infective agent that caused rabies was exposed, it would weaken and become less infective over time. He hypothesized that if that weakened material was then injected into a healthy animal, it might be able to survive a rabies infection. He developed this hypothesis by building upon his previous work with chickens, in which he discovered (by chance) that chickens infected with a naturally weakened form of cholera were later immune to a potent dose. Eventually, he tested his weakened rabies vaccine in live dogs, which he then infected with full-strength rabies. The dogs lived. The vaccine worked.

Joseph Meister

In 1865, Pasteur's vaccine was put to its most critical test when a young boy named Joseph Meister was mauled by a rabid dog. The boy was brought to Pasteur and, assuming that the boy would otherwise die of rabies, his mother gave Pasteur permission to administer the vaccine. As his first published attempt to use the vaccine in a human, it was a dramatic and risky attempt to save the boy's life. But it paid off. Pasteur had administered the vaccine over 11 days, and several months later Joseph Meister was not only alive but healthy; he would live until the 1940s.

What Pasteur didn't know was that the infective agent that causes rabies was impossibly small—a virus—and could not be directly observed at the time. So he had no other way, other than dogged trial-and-error experimentation and theorizing, to learn how to combat it. But because he put forth specific, falsifiable claims that were measurable (at least by the technology of the time), later scientists were able to re-examine his claims and results and build from them. Today, Pasteur is rightly considered one of the most important figures in the history of medicine and science, credited with greatly advancing germ theory and laying the groundwork for sanitation practices in hospitals and beyond.

66 Testing ideas by gathering evidence in a systematic manner is also known as the scientific method. 99

Pasteur's process may have seen variation over his career, but the path to a life-saving rabies vaccination followed the scientific method. A problem—rabies—was identified but had little to no explanation, so he used his past experience, his own work with microorganisms, and research done by his contemporaries to gather as much information as he could. Using what he'd learned, he came up with a hypothesis—a potential explanation and cure for the cause of rabies—and put it to the test through careful experimentation. From this, he was able to draw the conclusion that his vaccination worked. Through repetition, both by Pasteur and later scientists, more was learned about this process and the true nature of rabies and the virus that causes it.

THE PRODUCTS OF THE SCIENTIFIC METHOD: THEORIES, LAWS, AND FACTS

The repeated process of observation, hypothesis generation, and hypothesis testing ultimately leads to the production of a body of knowledge that serves to explain our surroundings and experiences. The knowledge that is gained from the scientific method can be broken down into three types of products. One product of the scientific method is either confirmation or rejection of a scientific theory. Unlike hypotheses, scientific theories are broader statements that explain many observations and events. Germ theory, for example, is a scientific theory that states that microscopic organisms, called germs, can invade bodily tissue and inflict harm by disrupting normal bodily functioning and causing disease. Germ theory falls into the category of a theory because it explains many causal origins of many different types of diseases, including different types of bacteria, viruses, and fungi. Germ theory thus explains a large set of problems, rather than a single problem.

A scientific law is another product that emerges from the scientific method. A law is a well-established hypothesis that has been rigorously and repeatedly tested and scrutinized under various conditions. Laws are more common in the physical sciences, such as in physics and geology. The law of conservation of energy, for example, states that energy can be neither created nor destroyed. Instead, it can only be transferred or transformed from one form of energy to another.

Last, facts are different from scientific laws, hypotheses, and theories. We often assume that facts mean that something is true with total certainty. This is not the case in science. When scientists speak of facts, they mean a claim that it would be unreasonable to disagree with given the current results of scientific research and evidence. An example of a scientific fact is the theory of evolution through natural selection. So far, no evidence has emerged that provides a better scientific explanation for how new species emerge and change over time.

It is important to emphasize that the scientific method and its products are not just meaningful to scientists. The scientific method impacts us in a number of different ways, from informing us about which medications we should take, the type of food we should eat, and how to best educate our children. Everyone can use the scientific method to help evaluate information and distinguish science from pseudoscience. Pseudoscientific claims rarely move beyond basic hypothesis generation. The so called "research" presented on websites pertaining to New Age metaphysics or in such books often consists of anecdotal claims and personal stories. But one person's isolated experience, by itself, is subject to bias, misinterpretation, simple wishful thinking, and all the normal varieties of human error.

The scientific method tries to eliminate the possibility of human error via repeated testing, worldwide replicability, and clearly stated claims. As Pasteur's famous example shows, science is by no means immune from error, but it instead learns from false claims, resulting in more-correct ones over time.

HOW TO DISTINGUISH SCIENCE FROM PSEUDOSCIENCE

SCIENCE STRIVES FOR FALSIFICATION RATHER THAN VERIFICATION

A good scientist isn't just interested in proving their own theory; in fact, negative results are equally interesting and can be just as useful as positive ones. Scientists must be open to the idea that their theory or hypothesis could be wrong, both now and in the future after more findings are made. But a negative result often leads to the formation of a revised, better hypothesis. Continual scrutiny of scientific ideas is how we ensure that science moves forward, building upon itself and improving our knowledge.

Compare that to the primary goal of crystal-healing practitioners, which is to convince you that crystals contain mystical energetic properties that can treat a plethora of ailments. To make a compelling case for why you should believe in crystal healing, self-proclaimed experts will present the "proof" of its effectiveness. As a general rule, you should always be wary of anyone claiming to definitely prove something's existence. They're usually more concerned with selling you something than they are with providing well-researched evidence for their claims.

In science, a disproved claim can be important, or even constitute a breakthrough. (Nicolaus Copernicus is a great example: he helped disprove the idea that the sun orbited the earth, the prevailing idea at the time). In contrast, peddlers of pseudoscience rarely present contradicting, or disconfirming evidence of their claims. Instead, they only present evidence that confirms their claims. Doing so is savvy, as it takes advantage of a vulnerability we all share—susceptibility to confirmation bias—a tendency to select or interpret information in a way that confirms the beliefs we already hold. Often occurring subconsciously, especially with deeply held beliefs, confirmation bias is a driving force behind the persistence of crystal healing and other metaphysical ideas. If so-called theories or hypotheses in the crystal-healing realm are ever scientifically tested, negative results are thrown out for one concocted reason or another (e.g., "the test wasn't performed properly"), while only tests yielding positive results are reported. Confirmation bias occurs outside of experimental settings as well; if a crystal healer's session does not yield relief for a patient, for example, the healer may try to explain that it wasn't because the healing was performed incorrectly, but because the patient wasn't open-minded enough to receive the healing energy, or perhaps that the patient requires more healing than one session can provide (and pay another fee). Or, conversely, if the session resulted in a positive reaction by the patient, that would be used as evidence that crystal healing works, for everyone, while ignoring the other times that sessions did not end in favorable reviews from patients.

SCIENTIFIC SIDEBAR

Confirmation Bias is the tendency to seek only information and evidence that supports what you currently believe in while ignoring and dismissing information that runs counter to your beliefs.

An emphasis on positive evidence stands in stark contrast to the scientific approach. Scientists search for information that disconfirms, or falsifies, their hypothesis or claims.

For an example from the history of science, consider the case of Nicolaus Copernicus again. In his groundbreaking book, *On the Revolutions of the Celestial Spheres*, published in 1543, he argued that the planets orbited the sun, challenging the prevailing theory of the time (which stated that the sun and other planets orbited Earth). Not all of the claims in his book were correct, including the claim that the planets orbited the sun in perfect circles, but nevertheless his book presented compelling new ideas that future scientists could build upon.

The flaws in the Copernican model were later corrected by astronomer and mathematician Johannes Kepler, who conducted his own research that utilized the careful, long-term celestial observations of his mentor, astronomer Tycho Brahe. As a result, astronomers were able to understand and predict the movement of the planets for the first time. By building upon the work of Copernicus and the observations of Brahe, Kepler was able to disconfirm certain prior claims and generate a new hypothesis that the orbit of the planets around the sun weren't perfectly round, but actually elliptical (slightly oval). He was right, and his discovery and the resulting math that backed it up (known as Kepler's Laws of Planetary Motion) helped lay the groundwork for Isaac Newton's physics and modern science in general.

SCIENTIFIC SIDEBAR

Extraordinary claims in the absence of extraordinary evidence:

Crystal-healing guides often have a habit of beginning somewhat scientifically, stating basic facts about rocks and minerals or geology as a whole. But they quickly veer into the pseudoscience realm, often incorporating scientific-sounding terms like "quantum," "infusions," and "matter-energy," without any explanation of what those terms mean. But this pseudoscientific gibberish is always presented confidently and in absolute terms, intended to be taken at face value and not to be questioned. After all, crystal-healing books are always written by people who seem to really know what they're talking about.

But instead of any credible evidence for the healing energies of crystals, what you'll typically find in small font at the beginning of a crystal healing book is a disclaimer that none of the information presented has been scientifically evaluated and shouldn't replace medical advice. That, in itself, should be a warning to the reader that the claims in the book are unsupported. So why do these authors always sound so confident? Because many of them have more to gain than just book sales. Many also sell gems and jewelry, teach crystal-healing classes, and/or offer services as alternative medicine "consultants" or "lifestyle coaches." In short, they don't need to offer carefully researched evidence to support their extraordinary claims because they're attempting to justify them in countless other ways.

SCIENCE THRIVES ON OPEN-MINDED SKEPTICISM INSTEAD OF ABSOLUTE CERTAINTY

Crystal-healing books and websites often present themselves in a scientific or all-knowing tone with titles such as "The Encyclopedia of Crystal Healing," or dogmatic titles like "The Crystal Bible." The intent of these statements is to give an air of authority and absolute certainty. This runs counter to the goals of science. According to noted astronomer and science communicator, Carl Sagan, science should operate as "open-minded skepticism." This means that rather than boasting confidence in facts and knowledge of the universe, scientists are open to new ideas and information that could discredit their current theories and beliefs. Scientists must accept that they could be getting some elements of their theory wrong, or may need to scrap the whole thing and start over. This is ultimately what leads to progress. In contrast to scientists, peddlers of pseudoscience fail to consider opposing information and instead gravitate toward blind faith in their ideas and claims.

It is no small matter that open-minded skepticism is a key mantra in science. This approach to the world runs counter to our desire for certainty and confirmation of our beliefs. Once a belief is formed, it is much harder to incorporate conflicting evidence and change it. Scientists must avoid blind devotion to their ideas and accept the possibility that new findings could contradict everything they currently know. This is why physicists, doctors, and other scientists are always careful to frame things in terms of probability or likelihood rather than in absolutes. That's why it's best to avoid the language of certainty, as it can discourage critical thinking and stifle our curiosity to search for other, better explanations. (Even the pioneering work in physics by Isaac Newton was eventually replaced by the improved work of Albert Einstein.)

SCIENCE DIFFERENTIATES CORRELATION FROM CAUSATION

Demonstrating cause and effect is a major component of scientific pursuits. Using the scientific method, a new treatment is put through rigorous testing to build an evidence base that supports its effectiveness and safety. Usually decades of research go into studying a new medicine or form of therapy, and most of the time, studies fail. Successful treatments are far rarer than failed attempts.

This level of rigor contrasts with the crystal-healing movement and other pseudoscientific practices. In pseudoscience, the so-called evidence is much weaker and less rigorous, usually based on anecdotal evidence (patient testimonials) from those who claim to have experienced a positive outcome.

But discerning between causation and correlation is difficult. Correlational evidence simply demonstrates that a relationship exists between two events. For example, a proponent of crystal healing may claim that when they meditate in front of rose quartz they are less likely to have headaches. Thus, there is a correlation between rose quartz and headaches. But simple correlation is not enough evidence to prove causation. As an example, consider the humorous website *Spurious Correlations*, created by Tyler Vigen, and the book of the same name; in it, Vigen compares unrelated real-world events that just so happen to be correlated. For instance, the number of letters in the winning word of the Scripps National Spelling Bee competition correlates rather closely with the number of people killed by venomous spiders in the United States over time. Obviously, spelling bee words have nothing to do with spider bites, but the correlation exists.

Correlational evidence therefore is not sufficient to claim a causal connection between rose quartz and headaches because there could be many other factors at work that are driving the presumed connection between the two. Correlation does not equal causation.

In order to claim a **causal** relationship, we need to be able to demonstrate how a treatment or relationship works. In the case of crystal healing, we'd need to know how rose quartz could prevent headaches. Other possible explanations, such as simple coincidence or the placebo effect (read more on page 112), must be ruled out via experiments in order to systematically test the effects of rose quartz on headache prevention in a group of people who volunteer to be studied.

In contrast, pseudoscientific evidence overstates correlational findings and misleads readers into believing that a causal relationship exists between crystals and a desired outcome without proper experimentation. Even when so-called experiments are described in crystal-healing books, they usually fail to meet strict scientific standards or have cherry-picked data.

SCIENCE SUCCEEDS VIA COLLABORATION, NOT IN ISOLATION

Pseudoscience thrives largely in isolation, particularly via the "definitive" nature of the topic's books and websites. There is little meaningful collaboration among pseudoscientists, and as a result fields like crystal healing are replete with individual self-proclaimed "experts," "teachers," or "masters" who often profess to have secret or ancient knowledge about the universe. In order to maintain their claims of expertise, practitioners of pseudoscience often disguise or distort the exact ingredients of methods of a treatment, or sometimes falsify it entirely. This lack of collaboration and consensus often results in conflicting information across pseudoscience books and other media.

In contrast, science is a collaborative process that aims to collectively advance our understanding and knowledge of the universe without claiming definitive, final answers to life's mysteries. Scientists strive to build connections with other researchers. Ultimately, scientists aim to publish their work so that it can reach the broader scientific community and public discourse. Publication of findings involves an extensive process called peer review. Peer reviewers are other knowledgeable scientists from the same research domain who deem whether the findings have merit.

Scientists also recognize and build off of the research conducted previously by other scientists. To make sure their work is understandable and reproducible, scientists clearly describe how they conducted their studies and how they reached the conclusions that they did. This is necessary for others in their field to replicate and test their findings under new conditions. Instead of acting as all-knowing enlightened "masters," scientists instead focus on specific domains of knowledge (a subset of biology, say, or a medical doctor specializing in a particular disease or related group of diseases). Along the way, they continually refine their practices and findings, encouraging others to do the same. This is why chemotherapy treatments, for example, are so different today than they were in the 1980s. A constant analysis and refinement of current methods has led to more effective treatments—and more survival—over time.

Types of evidence, ranked by levels of trustworthiness:

When reading about crystal healing from multiple sources, both for and against, you will come across different sources of evidence. Here we rank the level of trustworthiness of each type of claim from least (1) to greatest (5), followed by an example of such a claim.

1. Hearsay/Anecdotal evidence "A friend's mental and physical health improved after using crystals."

2. Case Study "After observing a cancer patient who meditated with crystals for a year, we found that his/her mental and physical health did not improve."

3. Survey "We surveyed 40 crystal-healing practitioners for a year and found that mental and physical health did improve"

4. Correlational Study "We compared the mental and physical health of practitioners and non-practitioners of crystal healing and found no difference between the two groups."

5. Experimental Study "We randomly assigned a group of participants to meditate with a real quartz crystal and another group to meditate with a fake quartz crystal that they believed was real. There was no difference in mental and physical health between the two groups."

Although experimental studies are ranked as most trustworthy, these too can also be flawed and should always be further scrutinized to check for biases in methodology and conclusions. Studies that can be replicated (repeated) offer the best evidence.

CHALLENGES TO OVERCOMING METAPHYSICAL AND PSEUDOSCIENTIFIC BELIEFS

Once beliefs about New Age metaphysics or crystal healing are entrenched, it can be difficult to change them. Many pseudoscientific claims are compelling because they prey on our emotions and vulnerabilities. At our core, we are predisposed to value and trust our own personal experiences more than the information that we gather secondhand. We also have a tendency to take our subjective experience as a sort of universal truth. In the context of crystal healing, this means that if one person believes they were healed by crystals, they then conclude that everyone can, or at least could, also be healed. Because of confirmation bias, anyone presenting opposing views may be disregarded as a disbeliever. And if their beliefs are strong enough, attempts to spread the ideas are likely to follow.

Another challenge is that pseudoscientific beliefs like crystal healing are rarely studied in a proper scientific sense. That's due in part to the fact that scientific resources are better used on more-promising therapeutics and treatments with more proven results. But it's also due to the fact that pseudoscience is inherently difficult to disprove. Within the scientific method, in order to test something there must be measurable and repeatable findings, but none exist for crystal healing and its invisible energy. This gives many crystal-healing practitioners license to make bold assumptions, such as "you can't prove crystal healing isn't real, so therefore it can be real." Perhaps more problematically, it also gives them cover to say, "prove to me that crystal healing doesn't work." But the burden of proof is always on the person making a claim. If you say, "crystals can help with diabetes," then you need to explain in specific, measurable terms how they work.

Instead, as adherents of pseudoscience are unable to provide concrete, verifiable facts to support their claims, they often resort to cobbling together an assemblage of scientific-sounding terms, cherry-picked research results, and out-of-context facts that seemingly support their metaphysics. All of this can come in conjunction with chastising scientists for being unable to falsify their wild claims. This is a pick-and-choose approach to reality that's indicative of deeply entrenched confirmation bias.

If you want to engage in a more critical and balanced approach to consuming information, then you can apply the set of practices encouraged by the scientific community. Scientists recognize that conclusions are not always iron-clad (and sometimes only temporary), so they put their trust in what has the most supporting information, for the moment. When you come across information that is supported by multiple, reputable sources, you can have more confidence in the message that is conveyed. Following this approach of open-minded skepticism can be beneficial in our everyday lives. When presented with information that opposes our currently held beliefs, instead of immediately discrediting it we should do our best to try and understand it and see how well it is supported. Additionally, we should try to engage with others who hold different beliefs from ours. This is not always easy. On the contrary, it's often hard. All too often we are caught in an echo chamber where we are surrounded by those who share our belief system. Combined, openness to new ideas and acceptance of uncertainty encourages curiosity and facilitates learning new information that can ultimately improve our lives.

The rare and beautiful blue "fuzz" of cyanotrichite crystals put it among the most endearing minerals.

Co-opting scientific terms to legitimize pseudoscientific claims:

While some crystal-healing authors openly admit their claims have no grounding in scientific evidence, other authors take the opposite approach by directly referencing scientific theories and principles in the hopes of legitimizing their claims. Many crystal-healing books alternate between generally accurate breakdowns of what a given mineral is, to providing all manner of misleading claims about what the crystals "do" for a given condition or affliction.

Many of the common buzzwords, such as the great emphasis placed on "vibrations," are easy to pick out, but other words like "piezoelectric" and "pyroelectric" are rooted in mineralogical science and refer to real properties of crystals. Suddenly, it becomes harder for the casual reader to separate fact from pseudoscience. Another increasingly common term appearing in crystal-healing books is "quantum mechanics," consistently used as evidence for why crystal healing is so difficult to explain. For the uninitiated, quantum mechanics is a deeply complex branch of physics concerned with the interaction of atoms and their subsequent particles. This field is difficult for even many scientists to fully understand, so attempting to use it to explain how crystal healing "works" to a layperson is essentially meaningless jargon.

It's not difficult to see how pseudoscience authors are able to put together a scientific-sounding rationale for why crystal healing supposedly works without even knowing themselves what the terms mean. By plucking impressive-sounding buzzwords from esoteric fields in which very few people have any expertise, they make their reader believe their claims are legitimate because the typical reader is ill-equipped to refute them. Even if those buzzwords have nothing to do with human health or the conditions they're used to "treat."

Gemmy yellow apatite crystals on their host rock from Mexico

Some of the most beautiful minerals are those that are too small to appreciate with the naked eye (and certainly too small to capture crystal healers' attention). These tiny cavansite crystal clusters measure only 3-5 millimeters each.

SCIENTIFIC SPOTLIGHT: COMPONENTS OF EXPERIMENTAL RESEARCH

A well-designed experiment is considered the gold standard for determining the safety and effectiveness of a treatment. High-quality experiments have three facets: experimental control, double-blind experimentation, and random assignment.

1. Experimental Control Experimental control means researchers do their best to isolate the true cause of an outcome and rule out extraneous factors that could bias the results. This is usually done by giving one group the hypothesized treatment and comparing the treatment group to a control group that lacks the therapeutic element. For example, In 2001, Dr. Christopher French conducted an experiment where he gave half of the participants a real crystal and the other half a cheap plastic crystal, unbeknownst to them, and asked both groups to meditate with the crystal and report their sensations. Both groups reported similar sensations, suggesting that the real crystal did not have any unique effects on participants when compared to participants from the control group.

2. Double-Blind Experimentation When possible, studies should implement double-blind experimentation. Double-blind means that both the researcher and the participants are unaware of who is in the treatment group and who is in the control group. This matters because researchers who know that a participant is receiving treatment are more likely to commit confirmation bias by giving more weight to evidence that confirms the hypothesis than disconfirms it. Participants who know or think they are in a treatment group may also act differently than if they are unaware, which could also bias results.

3. Random Assignment Scientists want to make sure that there are no differences between the control group and treatment group besides the treatment itself. Random assignment of participants to either the treatment or control group ensures that pre-existing differences are distributed throughout both groups. If assignment to groups was done non-randomly, like assigning all women to the control group and all men to the treatment group, then it would be difficult to conclude whether the differences were due to the treatment or due to differences in group selection.

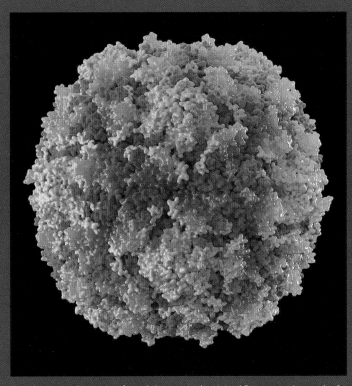

Eradicating polio, a once-feared disease, in the United States was the result of many carefully controlled studies and a great deal of research.

GUIDE TO SPOTTING PSEUDOSCIENCE

	Pseudoscience	Science
Types of claims made	• Makes extraordinary claims in the absence of evidence • States unfalsifiable claims that cannot be tested	• States rational claims that are framed in terms of probability or likelihood • Makes falsifiable claims that are backed by the scientific method • Acknowledges limitations
Types of evidence presented	• Relies on anecdotal evidence • Relies on personal experiences that are easy to overgeneralize • Treats correlation as causation	• Follows the scientific method of rigorous observation and experimentation • Distinguishes between correlational and causal evidence
Connections made to other evidence	• Makes assertions that are not grounded in previous research • Draws conclusions that run counter to established findings • Lacks self-correction: fails to take in new evidence • Relies on arguments from antiquity ("because it was practiced a thousand years ago, it must be true")	• Connects new concepts and ideas to previous research • Assertions logically follow from previous findings • Open to contradicting evidence that disconfirms previously held beliefs

	Pseudoscience	Science
Types of language used	• Uses scientific jargon to seem more legitimate • Makes dogmatic and all-knowing claims • Uses unclear or improperly defined terms	• Uses appropriate language and terminology • Defines terms • Provides necessary details • Avoids claims of absolute certainty
Character-istics of the community	• Findings are not evaluated systematically with peer review • Assertions are self-published and self-promoted by people/institutions who have a conflict of interest • Evidence presented cannot be independently tested and replicated	• Findings are peer-reviewed before they are published • Scientists are required to state if they have a conflict of interest • Assertions and conclusions are independently tested and replicated

CRYSTAL HEALING: CLAIMS AND REALITY

The various claims and techniques employed by crystal healers aren't often rooted in reality and can't stand up to scientific scrutiny, but nevertheless, some users claim that they've had success with these practices. What's more, they may even have real, measurable improvements in health and wellbeing after using crystals. We know that crystals don't actually "heal" anyone, so what's really going on? In this chapter we'll look at some of the practices crystal healers use and why they may appear to work for them.

"Crystals do have energy, but in the same way that a brick, a tree, or a glass of water does. The difference between the real energy that crystals contain and the energy purported by crystal healers is that crystals do not *emit* any energy."

HOW DOES CRYSTAL HEALING "WORK"?

When first investigating the claims of crystal-healing advocates, the first question that might come up is "What is crystal energy and where does it come from?" Proponents of metaphysics claim that all things in the universe are connected via energy, and that by being "open" to this energy and learning to perceive it, we can harness it to connect to elements of the universe outside of our human bodies. Think of it like a city's power grid—electric lines are all around you in a typical city, and with the proper knowledge and equipment, you can plug into it and turn on your lights—but you may not realize that most of a city's power lines are beneath your feet until you've been made aware of them. To crystal-healing proponents, this is why most people are not aware of the magical source of energy behind crystals—they have not been "awakened" to the energy that surrounds them. The energy they attribute to crystals therefore comes from "all around us."

Of course, this is not an inherently flawed concept—on a sunny day, for example, the sun's energy surrounds us in the form of light, infrared radiation, and other electromagnetic waves—but crystal-healing supporters have a difficult time explaining the physical source of the crystal "metaphysical energy." In many crystal healing books, the question is not even asked, it is merely posited as a fact that this energy exists. Others attribute it to vague "ancient origins" or unexplainable divine phenomena but offer no evidence. In fact, if you read ten crystal-healing books from different authors, you could find ten different proposed sources of metaphysical energy.

Such inconsistencies between published "experts" would be highly problematic in science, but it poses no problem to pseudoscientists. Disagreement and constructive criticism is encouraged in true scientific discourse, as difficult questions serve to discard inaccurate theories and help us better understand our world. But in New Age metaphysics, especially crystal healing, differences between writings come down to individual authors' experiences with crystals, each subjective opinion equally as valid—or invalid—as the next. This is well illustrated by looking at how crystal-healing authors discuss how crystal energy is transmitted. They offer several ideas, but two in particular tend to recur. The first states that crystals themselves contain energy, not unlike a battery, and can impart their energy but will eventually be depleted and require "recharging." And a second prominent claim states that crystals are actually just conduits for the earth's energy, acting as a way to focus the energy emitted by the earth toward the user. Both concepts are quite disparate yet both are largely accepted by crystal healers as possibilities.

Clearly the "we can have it all" approach to fact, and the inconsistent and sometimes contradictory nature of New Age metaphysical beliefs is problematic, but as previously discussed, it can be difficult to convince a believer that these flaws exist without firm evidence to the contrary. Luckily, the scientific principles behind geology and mineralogy, the energy they contain and how it can (or can't) be transmitted have long been understood.

DO CRYSTALS HAVE ENERGY?

Anything composed of atoms—which is to say, every object in the known universe—has energy. All matter does; this is a core fact of the natural world. The way in which atoms vibrate and electrically bond together to form molecules and crystals is a direct illustration of the innate energy that atoms hold. So of course crystals have energy, but in the same way that a brick, a tree, or a glass of water does. The difference between the real energy that crystals contain and the energy purported by crystal healers is that most crystals do not *emit* energy in any way. Quartz, for example, is the most common single mineral on Earth's surface and consists of just silicon and oxygen. Those two elements combine in such a way that their molecules are inert once crystallized, meaning that they do not change or emit energy in any way under normal conditions. In fact, quartz is among the most stable of all minerals, and it is a primary constituent of the earth's crust. One of the only ways quartz can emit energy is if a crystal is heated; it will continue to radiate heat (energy) until it has returned to room temperature, just as anything else would.

While the vast majority of minerals are stable and their crystals emit no energy, there actually are a few that do just the opposite, sending invisible, powerfully energetic particles into the world: radioactive minerals. You've probably heard of radioactive materials before; they contain unstable elements like uranium and thorium that cast off electrons in the form of radiation. But the majority of crystal healers won't recommend these minerals for healing practices because this kind of radiation (ionizing radiation) is inherently harmful to living tissue. As unstable elements project electrons, they are able to penetrate into our bodies and damage cells, potentially causing illnesses such as cancer. But since radioactive minerals are the only ones that actually emit energy, it's curious that crystal healers don't tell you to use them.

SCIENTIFIC SIDEBAR

Beyond their surface temperatures, our bodies cannot perceive the energy of typical solid objects, including crystals. Our senses are simply not sensitive enough to feel energy as slight as the latent energy in solid objects without an external stimulus, such as applying an electric current. If we were sensitive enough to feel the energy of the atoms within crystals, we'd also be able to feel the energy of all other things at all times, which would certainly be an overwhelming and likely unusable amount of information for our bodies to deal with. It would likely also be unpleasant.

Many crystal-healing books will tell you to steer clear of uraninite, thorite, carnotite, and the multitude of other often beautiful, but radioactive, minerals. This is because the authors know of the dangers of exposure to them. This should be a telling fact: if someone tries to convince you of invisible, undetectable energy emitted by harmless minerals while also warning you about the real and measurable harmful energy emitted by radioactive minerals, that should give you some indication of the credibility and intention of their claims.

To summarize: all crystals contain energy, because all matter is energetic. But beyond the physical temperature of a crystal, our bodies cannot perceive this latent energy any more than we can perceive it in water or pavement or a car tire. There are a few minerals that actually emit energy, but those that do are harmful (i.e., they can lead to cancer).

WHY CAN'T CRYSTALS EMIT ENERGY?

Radioactive minerals and their unstable elements aside, the reasons why typical minerals, such as quartz, can't emit energy are well understood. The two elements that comprise quartz—silicon and oxygen—are stable and will remain bonded indefinitely unless subjected to a chemical reaction. Absent such a reaction, we know that it doesn't emit anything because of a longtime scientific law: the Law of Conservation of Mass.

The notion of atoms as the core building block of matter dates back to as far as Democritus and the Ancient Greeks, but the Law of Conservation of Mass as we know it dates back to Antoine Lavoisier in the late eighteenth century. Put simply, the law states that mass is never created nor destroyed, but merely changes forms. Essentially, if a quartz crystal were emitting energy, it would be losing particles that have mass (such as electrons) to the surrounding environment. Those energetic particles

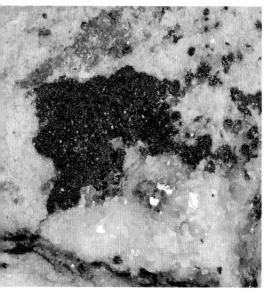

While it may look innocuous, the black mineral in this rock is uraninite, a highly radioactive mineral containing uranium.

would have to come from somewhere—from the elements comprising the crystal structure of the quartz—and therefore the crystal itself would be changed as a result. We see this in radioactive minerals. For example, uranium is a radioactive element. As uranium sheds electrons in a process called radioactive decay, the material left over decays into a different element. As uranium decays, it turns into helium and lead. Minerals containing uranium, such as uraninite, which is composed of uranium and oxygen, always also contain some amount of lead and trace helium as a result. Unless subjected to some outside chemical change, quartz always remains quartz, with no change to its crystal structure or elemental composition, therefore it does not emit energy. This is supported by repeated observations and measurements. If you doubt this, ask yourself: What *kind* of energy does it emit? Light? Radio waves? X-rays? Of course, crystal-healing advocates won't name the specific kind of energy for critics to test because it'd be simple to debunk.

Crystal healers often respond to such criticisms that the type of energy used in metaphysics is not the same kind of energy that we typically perceive, and when pressed, they don't provide much, if anything, more by way of evidence. You essentially have to take their word for it. If we allow that kind of logic into the discussion, then we are simply arguing the existence of magic. Without evidence, there is no arguing for or against such a baseless claim—it's pseudoscience in its most problematic form.

HOW IS CRYSTAL HEALING PERFORMED AND WHAT DOES IT REALLY DO?

To begin to understand the ways in which crystal healing is said to be beneficial, and the ways in which it actually could be beneficial (albeit often inadvertently) you need to know the basic ways crystal healing is performed. Crystal healing techniques, many appropriated from other cultures, vary a great deal. While some certainly see wider use and acceptance than others, the following are some of the most prominent examples.

LAYING ON THE BODY

WHAT HEALERS CLAIM

One of the most common and widespread methods of crystal healing is the simple laying of crystals and stones upon the body, particularly at joints, muscles, or other points of soreness or tension. This method involves the direct contact of crystals upon the skin, usually as the patient is lying still in a meditative state. Along with silence and deep breathing, the crystals are said to impart their energy into the patient. A healing session of this kind ends when the person receiving the energy feels satisfied, when the practitioner decides that the process has ended, or—believe it or not—sessions may also end when the crystals begin to fall off the body "of their own accord," indicating that

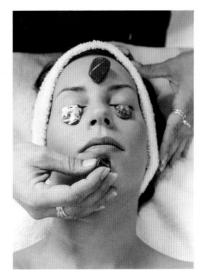

Different minerals are placed on various body parts for different intended effects. The stone on the woman's forehead is placed at the "third eye" chakra point.

they have finished their job. Sometimes massage is incorporated into this practice as well, both utilizing the stones as massaging tools—including stones shaped into wheel-like "crystal rollers," said to be particularly good for the skin—or simply by providing a massage in the presence of crystals.

This method comes largely from the notion of chakras, or energy points, which was appropriated from a number of religious and cultural practices, especially those from India. Now integrated in other cultures' practices as well, especially New Age systems of belief, this idea states that the human body contains five to seven chakras, which are the points at which our physical bodies intersect with our "subtle bodies," or internal spiritual selves. Placing crystals at the various chakra points is said to be a particularly powerful way to feel the crystals' energy. Keeping each chakra properly energized, not letting any get too weak or too powerful, is said to be the way to achieve bodily and spiritual balance.

WHAT IS ACTUALLY GOING ON

In most forms of laying crystals upon the body, great importance is put upon lying still, relaxing, remaining silent and breathing deeply, all while clearing your mind of worries and daily concerns. As it happens, these are all core tenants of meditation, which has repeatedly been shown to have measurable positive effects on your mind and body. The benefits of meditation can certainly be achieved

Stones may be used as massage tools and are said to impart energy directly to afflicted body parts.

without crystals, but for some, the addition of crystals may be the inspiration to actually follow through with the otherwise self-motivated and often difficult practice of regular meditation. Even without the pretense of meditation, simply lying silently and relaxing in the middle of the day can be a rejuvenating break from your normal routine and leave you feeling more energetic. But, as crystal healers and believers of New Age metaphysics are wont to do, these positive effects are inevitably attributed to the crystals, rather than to the very real benefits of relaxation and meditation.

CARRYING CHARMS, AMULETS, AND TOTEMS
WHAT HEALERS CLAIM

One of the most common and innocuous ways of incorporating crystals into your life is to carry or wear them as charms or amulets, or to keep carved stones in your home and office as totems. Beyond mere decoration, crystal-healing advocates purport that crystals keep you energized and mindful of your health as they impart energy into the room or directly into your body throughout the day. Healers claim that they do things like keep you motivated, increase your productivity or creativity,

Carrying stones, like this polished rose quartz, is claimed to impart their energies to you throughout the day.

or just generally output positive energy. Certain crystals carefully positioned around your home are also said to invite those crystals' attributes into your living space.

People have worn stones in their jewelry and kept them in their pockets and around their homes for millennia, with decorative purposes and superstitious beliefs attached in equal proportions. But believing that they are doing things like inviting wealth into your life or making you more outgoing is little more than wishful thinking, right? The truth is a little more complicated. Sometimes, by having a physical reminder—in this case, a stone or crystal—we can be more successful at enacting positive change in our lives. It's the same as the old method of tying a string around your finger as a reminder to do something later; the presence of the string, rather than the string itself, is meaningful. A type of crystal said to aid in controlling your appetite, for example, may actually seem to work when kept in your pocket because if you are constantly aware of its presence, you may be more self-motivated to eat less often. This is often referred to as a self-fulfilling prophecy. In short, because you want or hope that something will happen, you will both consciously and unconsciously make decisions that seem to make it happen—if it is within the realm of reality in the first place! The crystal itself isn't imparting any healing, but it can serve as a useful reminder.

> 66 The power of persuasion and the influence of a convincing personality are the keys to success in crystal healing. 99

Some grids are small, aided by drawn mandalas, and others are elaborate and large enough to sit within.

CRYSTAL GRIDS AND CRYSTAL BALANCING

WHAT HEALERS CLAIM

Crystal grids (also called layouts) are geometric arrangements of crystals and stones; users sit and meditate within a grid or in front of it. This method has increased in popularity in recent years, likely due to the more involved setup and number of stones required, which may feel to users as if they're doing more work and therefore should see more of a result. And by placing crystals at specific spots and in certain directions, sometimes facilitated by chanting or humming, grids are said to be a powerful way of gathering energy and imparting it to someone at its center. Sometimes a large crystal or carved crystal wand is used to "stimulate" the flow of energy. Users can also balance one or more crystals on their head, hands, or other parts of the body with the goal being to focus on nothing else but the balancing of the stones. As with laying crystals upon the body, the metaphysical idea at work is that by focusing on the grid and/or the balanced crystals you can receive mystical earthly energy.

Functionally identical in practice to laying crystals upon the body, crystal grids and crystal balancing are just extensions of the meditative concepts at work in many methods of crystal healing. By focusing on the present

A sunstone "wand" used to direct energy

moment and the crystals in the room, often with quiet music or silence, low light, and deep breathing, a crystal grid may promote a relaxed atmosphere that can leave a user feeling refreshed and with a clearer mind. The difference with grids is that the more elaborate preparation may seem to provide more of a "payoff." Taking the time to arrange a geometric pattern of crystals, often organized by size and color, is in itself a meditative process that can feel very important to the user. Meditating at the center of a carefully crafted grid may then feel that much more rewarding, and at the end of a session the user may be more apt to feel that the process was particularly successful. But as with other aspects of metaphysics, attributing a perceived health benefit to the crystals—rather than the relaxing and meditative effects of such a tactile, personal process as grid construction—is incorrect.

INFUSIONS AND BATHING
WHAT HEALERS CLAIM

The desire to attain a deeper connection to Earth and its energy through the use of rocks and minerals has led some crystal healers to take their beliefs a potentially dangerous step further by seeking a means of physically consuming crystals. The practices of producing infused water and mineral elixirs have been developed to fill the role. Infusions, in which "crystal essences" are imparted into water, are made by placing stones in purified water and allowing them to soak, usually in the sun or moonlight, for several hours. When the user drinks the water, they believe they are also consuming the metaphysical attributes of the crystals, which will act upon them from within. Also increasing in popularity are "crystal straws," which can either be made from minerals or are plastic or glass straws studded with adhered crystals, said to impart their energy to the water as you drink it. Mineral elixirs, on the other hand, are made from crushed and powdered minerals stirred into water and then consumed—a much more dramatic and potentially poisonous means to achieve what users believe are more potent results than those provided by infusions.

Mineral specimens soaking in water as an infusion is prepared

Similarly, bathing with crystals is said to be a means of literally "steeping" yourself in crystal energy. The water of a hot bath full of crystals and stones is believed to allow the energies to flow around and soak into your body.

WHAT IS ACTUALLY GOING ON

The desire to practice crystal healing in a more dramatic fashion has led some to attempt this questionable consumption of stones. People who regularly drink infusions, for example, may report very real feelings of better health. But as with most "results" of crystal healing, this can be attributed to something other than the crystals used. In this case, it is the water. Water is essential to our health, and increased water intake can make us feel more energetic, help our bodies more quickly recover from illness or simple maladies such as headaches, and promote more regular digestion. Studies find that the majority of people do not drink enough water throughout an average day, so increasing this intake can lead to measurable and tangible positive effects.

It may be beautiful, but drinking an infusion made from chalcanthite could be potentially fatal.

But adding crystals to water doesn't help. In many cases, most common crystal-healing minerals, such as quartz, are inert and would likely impart nothing to the water. Nothing is moving from the quartz into the water, except for imaginary healing energy.

Nonetheless, drinking mineral-infused water is unwise and should be avoided. Many minerals and crystals contain impurities, and without a proper knowledge of the compounds or elements present in a crystal or mineral specimen, some could be poisonous when soaked in water. Other common healing minerals, such as selenite, will dissolve in water and even if the compounds within aren't necessarily toxic, they could contain impurities that are. Others could be toxic in concentration, such as with halite or carnallite, both of which would dissolve completely. And of course there are minerals that would dissolve completely and are outright toxic, such as chalcanthite (which would turn the water a pretty blue color, but could kill you).

Mineral elixirs should also be avoided entirely. By grinding up a mineral, mixing the powder with water, and drinking it, you could be introducing toxic amounts of metals or other compounds that may not dissolve in the water but can still be harmful if consumed.

Lastly, regarding mineral baths—who doesn't enjoy a long soak in a hot bubble bath? For many people, this is an innately relaxing process, but next time, skip the crystals, as they don't add any benefits. In fact, hard crystals add potential discomfort, can lead to clogged drains, and you might wreck your favorite crystal as many are susceptible to dissolving or damage when immersed.

A quartz pendulum being used during a crystal therapy session to facilitate the flow of energy

PENDULUMS, SCRYING, AND GAZING

WHAT HEALERS CLAIM

With pendulums, scrying, and crystal gazing, the use of crystals moves beyond simple healing and attempts to look deeper into the metaphysical realm by attempting to contact and utilize spiritual forces. These activities are largely concerned with communication with spirits, often including the dead, and with receiving glimpses of the past or future. The use of crystal pendulums, which consist of a pointed crystal or carved stone at the end of a chain or string, is a form of dowsing, a pseudoscientific method of divining the location of a resource. In this case, the pendulum is held suspended, sometimes over a drawing or diagram, and begins to move in circular or semicircular motions, supposedly in response to the user's question. The motions are then interpreted in various ways to represent answers provided by the universe.

Scrying and gazing are two ways in which more direct answers are sought from the natural world. By peering into crystals, scryers claim to be able to receive messages from spiritual sources, while gazers are presented with dream-like images of the past, present, and future, as well as prophetic visions for them to interpret. In each case, crystals are seen as the link between our world and the supernatural world in which these messages originate. The natural facets of a crystal are said to amplify the availability and potency of these messages, as the natural angles and faces of a crystal are claimed to receive and transmit this information.

WHAT IS ACTUALLY GOING ON

The use of pendulums, crystal scrying, and crystal gazing is more akin to fortune telling than healing, and is therefore beyond the healing-oriented focus of this book. The results of such activities are entirely subjective, untestable, unverifiable, and nearly entirely imagined, but by using them to tell a compelling story to a willing subject, practitioners are able to perpetuate such fictions.

Nearly all of these practices are built upon the trust between the practitioner and their audience. A pendulum user, for example, will insist that they are holding their hand still while the pendulum begins to move on its own, but in so many cases (and in countless videos available on

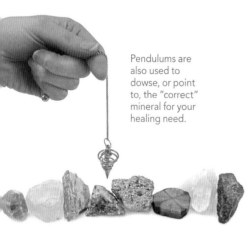

Pendulums are also used to dowse, or point to, the "correct" mineral for your healing need.

the internet), it is obvious to any objective viewer that their hand is indeed moving. Yet the practice continues to attract an audience. Clearly the power of persuasion and the influence of a convincing personality are the ultimate keys to success in such practices.

MAGNETS

WHAT HEALERS CLAIM

If there is one natural material that seems magical,
at least at first glance, it is
magnets. The attributes
of magnets and magnetic
minerals to attract and
be attracted to iron sim-
ply due to their inherent
crystal structure makes
them unique in the mineral
world. It is curious then
why many crystal-heal-

Magnetic bracelets of all styles are widely available on the
market, often with misleading claims attached.

ing authors place little added emphasis on magnets, especially since
lodestone, a magnetized form of the mineral magnetite, occurs naturally. Still, segments of the alternative medicine industry have latched
onto the special properties of magnets and asserted their own claims.
Magnetic bracelets, charms, or beads are common in the marketplace and
are said to promote circulation and healing in the joint or part of the body
on which it is worn due to an interaction with the iron in our blood.

WHAT IS ACTUALLY GOING ON

The concept that a magnetic bracelet will somehow attract more blood to
an arthritic wrist and ease its pain is based upon a flawed understanding
of how magnets actually affect the human body. While blood does contain
hemoglobin, an iron-bearing compound, no magnet as simple as those
in a bracelet could possibly be strong enough to affect it. In addition,
it is deoxygenated blood that tests have shown is very weakly attracted
to a strong magnet—this is the blood that has done its job of bringing

oxygen to our body parts and is on its way back to the heart. Oxygenated blood, which delivers life-giving oxygen to our bodies, is actually very weakly *repelled* by a strong magnet. Therefore, if our blood could possibly be affected by a piece of magnetic jewelry, it would be the "spent" blood that would be attracted to it, which would collect dangerously in the spot around the magnet. In short, it's a good thing that magnetic jewelry doesn't actually work. Yet people still buy and wear magnetic jewelry and purport benefits that are at their core are derived from wishful thinking and misleading marketing.

GEOPHAGIA (EATING EARTH)

WHAT HEALERS CLAIM

Easily the most fringe activity presented here, one only loosely related to crystal healing, geophagia, or the act of eating earth, has worked its way into the metaphysics and pseudoscience discussion. Popularized recently by some Hollywood actors, the act of consuming earth, usually in the form of dirt and clay, is said to have "grounding" and "detoxifying" effects, attuning the consumer's body to our planet, among other baseless claims. Practiced in the Western world only by the most die-hard few, geophagia has historically cropped up in many cultures and continues to occasionally do so, particularly as a reaction to famine and starvation.

WHAT IS ACTUALLY GOING ON

The search for individuality and the desire to push the envelope in alternative lifestyle practices has led to this potentially deadly practice appearing in modern culture. Aside from the gut-clogging nature of indigestible soil and dirt, most types of clay can absorb many times their weight in water, leading to intestinal blockages and other serious complications. In addition, soils can contain harmful bacteria, heavy metals, and pollution, even if obtained from a seemingly remote, pristine source. Ironically, soil eaten for its "detoxifying" effects may actually introduce toxic substances to the body.

66 The desire to practice crystal healing in a more dramatic fashion has led some to attempt this questionable and dangerous consumption of stones. 99

Clays can absorb many times their weight in water; if consumed, they can expand in your gut and create potentially deadly blockages.

Assortments of tumble-polished stones like this are common in shops, often with various positive attributes associated to each species.

INTANGIBLE ATTRIBUTES OF MINERALS

Each of the above crystal-healing practices can be said to treat any number of maladies and afflictions. One user may try meditating with a crystal grid to treat their chronic pain, while another may use the same technique to achieve creative inspiration. These very different uses "work" because the attributes of each different stone or crystal are said to vary greatly, and performing these techniques with different stones can be said to yield different results. In this way, each method of crystal healing can be highly tailored and personalized specifically to each user, which presents to many an appealing alternative to modern medicine when faith in science has waned.

These intangible but invariably positive attributes are said to be inherent to each type of rock or mineral, a reflection of their individual energy and "purpose" in the universe. Most rock and mineral species have many attributes attached, including some that are medicinal, such as "aids in circulation" or "helps with indigestion," and many more that are aspirational, such as "increases creativity" or "invites love into your life." In this way, specimens can be sold for all manners of metaphysical use, whether it be in a crystal grid, beaded "healing" jewelry, or as ornamentation around the home. These attributes are highly variable, often with little consensus among authors as to what a mineral may be "good for," and sometimes the information presented about a particular species in one book or blog may be completely contradicted by claims in another. But regardless of the endless inconsistencies, crystals' attributes are always presented as valid, with crystal-healing purveyors mostly untroubled by such conflicts because of crystal healing's overly positive, "make of it as you wish" outlook.

A larger quartz crystal "recharging" smaller crystals

CLEANING, PURIFYING, AND CHARGING CRYSTALS

For crystals to maintain their effectiveness in healing, most crystal healing authors and gurus will insist that crystals must be properly cleaned and purified, and then recharged before and/or after they've been used. Cleaning crystals is usually a simple matter of washing them in water, using soap if necessary, and drying them fully. Supposedly this does not interfere with their natural abilities. Purifying, however, is a more complicated matter. It is said that crystals must be purified of any negative energies in order for them to be useful for providing good healing energy. Authors often present varying ways purification can be done, usually including using other minerals known for their purifying abilities or burying an "impure" crystal in salt or other supposedly cleansing substance. After a sufficient period of time, sometimes in conjunction with a cleansing ritual of some sort or by exposure to sun or moonlight, the freshly cleansed crystals are said to then be ready to accept a charge.

Recharging crystals is rooted in the concept that crystals of greater size and clarity are "mother crystals," or crystals whose charge cannot run out and can be transmitted to lesser crystals in order to restore their curative properties. For whatever reason that these large and often very expensive specimens are said to contain infinite energy, they are encircled by smaller discharged crystals, often in a natural setting, and left to do their "work" of restoring the healing stones. Of course, no energy is actually passed between crystals or stones, but this extra ritualistic step likely makes crystal users feel like their crystal use is more effective.

Curiously, while emphasis is usually placed on cleaning crystals and keeping them free of impurities before use, many authors will say that it is OK to tape or otherwise adhere them to your body if they won't stay put. This creates an interesting double-standard that these authors choose to ignore in favor of providing an easy solution.

MISATTRIBUTION OF BENEFITS

The core concepts of crystal healing are relatively easy to debunk, so why do many people continue to practice it and insist that it works? In many respects, crystal healing and other metaphysical beliefs have become something like a faith, surviving in part due to a strong emphasis simply on belief. But beyond this sort of faith, practitioners claim very real benefits that they attest to strongly, benefits they attribute to their use of crystals, rocks, and minerals. So what leads to this belief?

In many cases, these feelings of success with crystal healing are due to the user giving New Age metaphysics undue credit for the healing or relief they feel. This misattribution of benefits is a driving force in the persistence of crystal healing, and in many other forms of pseudoscience. It's the same universal human flaw we find at work in superstition. Due to our desire to influence matters out of our direct control, we attribute good results with things that we can control, no matter how small. For example, when a sports fan wears their "lucky socks" to help their favorite team win. Of course the socks aren't doing anything, but the fan's feelings of efficacy and power in taking that small action lead to very real positive responses in the fan's mind when their team wins. In crystal healing and other metaphysical practices, the users feel a similar sense of control over things innately uncontrollable, and since they feel as though their effort should be meaningful, they are more apt to find positive results, whether real or imagined.

In reality, any real positive benefits derived from crystal healing are often easy to explain once the shroud of metaphysics has been removed. Returning to a previous example, certain stones are claimed to aid in recovery from the common cold. But the common cold is also easily fought off by your body's immune system, usually in just a few days in an otherwise healthy person. If your body happens to recover from the illness around the same time you tried crystal healing, you may be tempted to attribute

SCIENTIFIC SIDEBAR

Occam's Razor is a principle that states that the more speculation or assumptions you have to make in order to try to explain something, the more unlikely that explanation is. This is a philosophical principle known as Occam's Razor, which states that the simplest explanation is often the correct explanation.

the recovery to the stones. But in this and in countless other cases, the simplest explanation is often the correct one. Attributing benefits to needlessly complex and unproven, inexplicable causes is something we all do, due to the ways our brains work. But it is something that you can learn to overcome through critical thinking and evidence-based reasoning.

Specific examples of how crystal healers misattribute perceived benefits are plentiful and make it easy to see how a desire to affect one's health can cloud judgment and assessment of how or why crystals did or didn't work. In some cases, there may not have been any positive benefit at all, instead just the implication or suggestion of one, which in turn made it feel real. This is known as the placebo effect (for more, see page 112).

THE PROBLEM OF WHY IT "WORKS"

The ways in which crystal healing is claimed to work and the ways in which it may actually work are two different discussions—one rooted in the myths of metaphysics and the other in the science of human psychology. No matter your side in the argument, it is unfair to say that the practice of crystal healing is without any positive merit; some of the methods used in crystal healing (meditation) actually have real effects, but the cause of any positive attributes derived from the practice are invariably misattributed by believers. The benefits that may exist don't stem from earthly energy or magic, but instead from meditation, relaxation and deep breathing, or placebo, or in some cases, outright cognitive biases. People make flawed attributions like this in large part because our brain structure is essentially hardwired to "accept" some biases (for more, see page 102).

Crystal-healing practices also seem to "work" because the language and concepts that surround crystal healing are ambiguous and subjective, meaning different things to different users. There is virtually no right or wrong way to perform these methods. Crystal healing also survives in part due to the overwhelmingly positive and optimistic view and tone, which is cultivated by authors in the subject. By making readers and would-be practitioners comfortable, providing well-worded and self-important explanations of how metaphysics work, they make crystal healing feel like a safe, smart, and exciting alternative. They attempt to legitimize their claims by hijacking scientific principles and bending them to their own version of understanding, such as the difficult concepts at work in quantum physics or even quartz clocks. But as soon as you begin asking difficult questions of New Age metaphysics, then the excuses begin to flow.

THE PSYCHOLOGY OF CRYSTAL HEALING

Crystal-healing practices can activate many different feelings in users—satisfaction, pleasure, efficacy—but what's really going on in peoples' minds when crystals appear to work? As humans, we naturally seek out information that confirms beliefs we already hold, we make connections that don't really exist, and we may even feel positive effects from a treatment that wasn't real to begin with. These are all innate ways in which our minds work. In this chapter, we'll take a look at the science behind it.

"Over half of the U.S. population believes in some form of alternative healing, superstition, or psychic power. Psychological principles can help us explain why these beliefs are so prominent."

WHY CRYSTAL HEALING SEEMS TO "WORK"

Crystals aren't endowed with powers that can heal, so why do so many people believe in crystal healing? Understanding people's beliefs and behaviors falls within the realm of psychology, the scientific study of how people think and behave. Psychological principles can help us explain why crystal healing draws in new consumers and maintains such a loyal following.

First, it's important to note that pseudoscientific belief in general is such a widespread phenomenon. Over half of the U.S. population believes in some form of alternative healing, superstition, or psychic power. Notably, today's pseudoscientific beliefs do not seem to emerge from lack of intelligence or gullibility, or from mistrust in science or medicine. Even highly regarded professionals and past United States presidents have relied on astrology, telepathy or superstition to inform their decisions.

At the same time, science and medicine are still highly valued in America, despite what we may hear about the spread of anti-science beliefs. Reassuringly, a 2018 report by the American Academy of the Arts and Sciences revealed that a majority of citizens support science and are confident in the scientific community.

Puzzlingly, there are many people who simultaneously value science while also holding superstitious or pseudoscientific beliefs. To explain this contradiction, we need a better understanding of how we perceive the world and draw connections between cause and effect. In other words, how do we determine that something is real versus fake, or true versus false? These are questions that particularly concern cognitive psychologists, scientists that study the way we process and retain information and make decisions.

For a long time, cognitive psychologists theorized that people made decisions much like a computer processes information. In other words, information was thought to be processed in a rational, systematic way to minimize errors. But later scientific studies revealed that our decisions are not always so carefully calculated and we sometimes make mistakes. This is because on a daily basis we are bombarded with information and must make decisions on the fly. Life would be quite tedious if we had to systematically weigh the pros and cons of every choice. To enable faster reaction time and to solve problems more efficiently, we have developed mental tools—shortcuts, in a sense—to aid the decision-making process. One tool that we use is **pattern detection**. From an early age, we detect patterns in our environment that allow us to determine who and what will help us survive. Facial recognition and language development are two forms of pattern detection that emerge early on in our lives. Infants are capable of recognizing faces and identifying which noises coincide with specific objects or actions.

Causal inference is a type of thought process that allows us to detect patterns and make predictions about the future. When two events or actions coincide with each other, we have a tendency to assume that they are related. For example, if the mailman comes to your door and then your dog barks, you might then believe that your dog barked because of the mailman. The more common either of the two events are, and the more often they coincide with each other, the more likely you are to draw a causal connection. If you are at your house every time the mailman comes, and your dog frequently barks, you will likely assume that the mailman causes your dog to bark. But if your dog rarely barks, or if you are absent from your house when the mailman stops by, you may never make that causal connection.

Although pattern detection and causal inference are useful and make life more predictable, sometimes we assume there is a causal connection when in fact none exists. The incorrect assumption of a causal connection is called a **causal illusion**. We are all vulnerable to causal illusions because our emotions and desires bias how we perceive and interpret incoming information. Causal illusions can occur when we misattribute an outcome to the wrong cause based on our beliefs and desires. Let's return to the dog and mailman example. Because we want to believe that we have a well-behaved dog, we may be more motivated to attribute the cause of his barking to the mailman's intimidating stature, rather than attributing it to a flaw in the dog's training, which may be the more likely reason. In regard to crystal healing and susceptibility to causal illusions, when we are feeling sick or in pain, or

Vanadinite's crystals are often colorful and well-formed, with perfect hexagonal shapes. But what causes someone to believe that they'll help with the health of your lungs, as many crystal healers purport?

lonely, or just want to feel more in control of our hectic lives, we are motivated to get rid of what is troubling us. This motivation to feel better could color how we interpret information about crystal healing and increase our susceptibility to form causal illusions.

The self-help and alternative medicine industry draws on our desires and motivations for optimal health and self-improvement. Self-proclaimed experts bombard us with information to try and convince us that a particular method or practice will solve our problems. With the advent of social media, we are increasingly exposed to unregulated information and false claims that may lead us down the path toward causal illusions and pseudoscientific thinking, without realizing that we are being misled. Even before practicing crystal-healing ourselves, we may come to erroneously believe that it is an effective therapeutic practice.

Much of the information conveyed on crystal healing websites and books capitalizes on our susceptibility to form causal illusions. Persuasive patient testimonials and anecdotal evidence powerfully draw the connection between crystals (the proposed cause) and pain reduction or increased sense of empowerment (the proposed effects). Recall that we are more likely to form a causal connection when the coinciding causes and effects are common. Now, more than ever, crystals and minerals are widely marketed and made accessible through online and in-store sellers. Gurus will claim that crystals can heal back pain and headaches, which are common ailments, or can help us with our social and love lives, which many of us have difficulties with now and then. If we come to believe that many people are practicing crystal healing and that there seems to be evidence that they are effective—we might then be misled into attributing the desired outcomes to crystals.

This type of image, common in crystal healing books and websites, shows crystal use in an almost clinical presentation, aiding in its appearance of being a legitimate area of expertise.

The testimonials offered in books and blogs may be true: a couple may have found everlasting love at a crystal-healing workshop, or someone's back pain might have resolved after using crystals, but the role of crystals in bringing about those outcomes is likely just coincidental. Research shows that back pain and headaches can subside over time without any form of therapeutic intervention. Interpersonal issues tend to get resolved with patience and good communication. That is, there are much more likely alternative explanations that we should first rule out.

Thinking of alternative explanations that go against established beliefs is difficult to do. This is because practitioners of crystal healing often engage in what is known as confirmation bias. **Confirmation bias** is the tendency to consider only the information that is consistent with our beliefs and to ignore information that goes against firmly held beliefs. A website or book on crystal healing presents only the claims that tout the benefits, not any of the downsides or the instances where nothing of note happened after using crystals. Often the claims are unscientific,

meaning they are untestable and unfalsifiable, even if they are supposedly evidence-based. For example, because the "energy" that crystal healers describe is so ill-defined, it is not possible to detect the "energy" emitted from crystals or to measure how the body is healed from crystals directly.

Continuous exposure to one-sided information can lead to what is known as the **bias blind spot**. That is, advocates of crystal healing don't accept that their own beliefs may be biased while at the same time they believe that others' thinking is incorrect and biased. In other words, believers think that they are the experts and that non-believers are the ones who are naive and misguided. Believers often discredit opposing information because it is viewed as an exception or is flawed in some way. Ultimately, confirmation bias and the bias blind spot maintain the causal illusion of the connection between crystals and healing. The lesson here is that it's easy to place far too much faith in anecdotal evidence. One's own experiences and beliefs can absolutely seem more appealing than scientific evidence, especially if the science refutes one's closely held beliefs. But that doesn't mean your experiences (and the conclusions you draw from them) are universal, or even correct. For that, we need science.

In order to combat causal illusions, cognitive science offers us some helpful strategies. One of the best ways to counteract pseudoscientific beliefs is to expose believers to instances where the desired outcome was achieved in the absence of the attributed cause. Sure, holding onto a crystal may coincide with fewer headaches, but what about all the other times you didn't get a headache when you weren't holding a crystal? Another approach is to think of alternative explanations. For example, an elementary school teacher may incorrectly attribute her crystal ring to producing a calmer, more respectful classroom. But if she paid attention to when she wore her ring, she might notice that it coincided with the days that her class had morning gym and so students had a chance to burn off their extra energy. Want to know for certain if wearing a crystal necklace cures pain? This will require well-controlled scientific experiments, which are all too scarce in the alternative-healing realm. You would need to randomly assign half of a group of pain sufferers to wear a necklace made of genuine crystals and the other half to wear a necklace made of seemingly identical, but fake, crystals. If both real crystals and fake crystals produce the same results, it's highly unlikely the "energy" from crystals is what is driving the desired outcome.

Reflecting on historical events can also help us understand and combat pseudoscientific thinking. Back in the 1700s, Benjamin Franklin and a team of scientists discredited one of the world's original alternative healing gurus, Franz Mesmer (where the word "mesmerizing" originates). Mesmer was a German physician who claimed to have discovered a natural

A period etching showing Mesmer at work with his "animal magnetizer" vat and his willing patients

energy fluid that flowed within and between all animate beings and inanimate objects. This fluid was called "animal magnetism." He theorized that illness resulted from obstructions to the flow of fluid within us and that this obstruction could be cured through a specific ritual. In Paris, France, Mesmer set up a clinic where he performed these rituals in which he would pass his hands over the patient's body in a calming fashion and gently press below the diaphragm. Many patients claimed to be cured of their ailments and were satisfied with the results. Over time his clientele grew so large that Mesmer could only accommodate the demand by holding group treatments. Patients would stand together around a vat and grasp on to metal rods he called magnetizers. Mesmer claimed that the magnetizers were conduits for the flow of animal magnetism. He would then perform actions to restore the flow of energy within his patients and thus supposedly cure them of their illnesses.

Mesmer's treatments were costly, and as skepticism over their effectiveness grew, the King of France recruited Benjamin Franklin and other scientists to test whether Mesmer had indeed discovered a new physical fluid that was transmitted via the magnetizers. The scientists conducted a series of experiments. In one experiment, they replaced the magnetizers with fake ones of similar appearance and compared their effects to the real magnetizers. Lo and behold, the fake magnetizers produced the same results as the real ones, even though they were made of entirely different physical properties. This finding discredited Mesmer's claim that the flow of fluid through the magnetizers played a vital role in restoring the patients' health. So why did so many patients claim they were healed and thus maintain their loyalty to Mesmer? The scientists concluded that because Mesmer's claim was wrong, the effectiveness of the treatment comes down to the patient's belief rather than the actual physical properties of the treatment itself. This was one of the first double-blind studies conducted, and it was also an early demonstration of what we know as the placebo effect (see page 112).

An 1885 poster advertising a "mesmerism" demonstration, presented as "amusing and instructive" entertainment, even after Franklin had debunked it

Similar to Mesmer and his magnetizers, the belief that crystal healing is an effective form of therapy may in itself produce changes in the mind and body that lead to desirable outcomes in patients. Human motivations and emotions are psychological experiences that can impact our biology and the way we feel mentally and physically. Health psychology, a relatively new field of study, focuses on how our health is shaped by the combination of biological, psychological, and social processes. In other words, our health is not only a function of biology, but it is also a function of our social environment and how we think about and react to that environment. Consider how sharing a laugh with

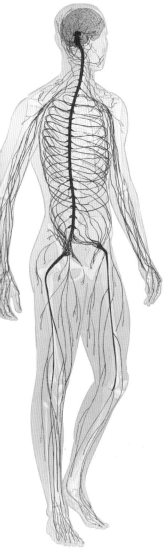

Our bodies' nervous system connects the brain to the rest of the body; the ways we think can affect the rest of our body by sending signals along these pathways.

friends makes you feel happy, lively, and energetic, or how a traffic jam fills you with frustration as your heart beats faster and your blood pressure spikes. These are just some illustrative examples of how our psychological experiences (thoughts and emotions) correspond with biological changes in our body in response to what is happening in our social world. Understanding the connections between the mind, body, and the social world can help to explain why alternative practices, like crystal healing, produce positive outcomes despite the lack of evidence-based therapeutic properties. Crystal healing may actually "work," but in a much different way than advertised.

THE PLACEBO EFFECT

In the medical field, to determine whether a medication or therapy is an effective treatment, it is compared to a placebo. A **placebo** is as similar to the medication or therapy as possible but lacks the theorized active ingredient. When testing a drug, for example, the placebo will mimic the drug in every way, except that it is an inert sugar pill. If the drug outperforms the placebo, then it is deemed to have a therapeutic active ingredient. So far, there is no evidence that crystal healing or other types of alternative practices outperform placebos.

Interestingly, however, some individuals undergoing the placebo treatment show improvements in mental and physical health conditions when compared to no treatment at all. This so-called **placebo effect** is a phenomenon of particular interest to health psychologists because it illustrates how much health is connected to beliefs, or psychological experiences. Scientific studies show that patients given a sugar pill (placebo) and told by a confident and positive doctor that it is an effective medicine are more likely to report improvement in physical health compared to patients who are not given any treatment. Remarkably, even patients who are aware they received a placebo still experience a reduction in their physical symptoms. Thus, deception does not seem to be a crucial factor.

Extensive research on the placebo effect illustrates why in some cases crystal healing could produce desirable outcomes. Rigorous scientific studies have broken down the three key factors of the placebo effect that lead to improvements in health: (1) the patient's response to being observed and assessed (known as the Hawthorne effect), (2) the patient's response to the placebo therapy or medicine, and (3) the patient's response to interacting with the practitioner. These three different components of the placebo effect may also apply to crystal healing.

1. Patient's response to being observed and assessed

The **Hawthorne effect** is what occurs when patients change their behavior as a result of being observed. For example, someone whose health is regularly monitored in a study with frequent clinic visits may begin to smoke and drink less, or exercise more because they know they are being monitored. This change in behavior, rather than the treatment itself, could be what is improving health. Likewise, people who regularly visit a crystal therapist may change their behavior in some way because they want to demonstrate they are progressing though the treatment. Being held accountable by another individual can be a powerful motivator to behave in a more socially desirable and healthier way. Carrying around a stone could also help you engage in more self-reflection, thus enhancing awareness of your behaviors. Increased awareness could lead you to change problematic behaviors, like sitting for too long or holding onto tension, by replacing them with more positive behaviors, like taking more walks and engaging in breathing and meditation exercises. This motivation and positive change in behavior, rather than the properties of crystals, could be plausible ways that crystal healing produces positive outcomes.

2. Patient's response to placebo treatment Some crystal-healing practices could also directly affect our biological functioning. Taking a moment to meditate with crystals, for example, could have calming effects on our body that makes us feel mentally and physically better. Indeed, our physical state is in part determined by our mental state. Communication between our brain and other organs (heart, stomach, intestines) is coordinated by chemical messengers called neurotransmitters and hormones.

To envision this, imagine the brain as a car, and it activates certain chemical messengers by stepping on either the gas pedal or the brake pedal. When the brain steps on the gas pedal, it sends specific messengers that tell other organs to speed up and burn energy. One way that this affects the body is that our heart beats faster, and we increase energy output. When the brakes are applied, it sends another set of messengers that tell the body to take it easy and conserve energy, and to slow down our heart rate.

Our social world and our emotions play a role in how much our brain steps on the gas pedal versus the brake pedal. When we face challenges, such as a work deadline or intense soccer match, this leads to heightened stress (the gas pedal) which helps us deal effectively with the challenge. However, if we experience uncontrollable and unpredictable challenges, we could become overwhelmed and experience a state of chronic stress. Chronic stress can lead to overactivation of the stress response, and over-exposure to stress hormones throughout the body. Overexposure to stress hormones could disrupt many biological functions, leading to impaired digestion, increases in our heart rate and blood pressure at inappropriate times, and disrupted sleep. Stress also increases tension throughout the body, contributing to the onset or worsening of pain. In the long run, both our mental and physical state can suffer as a result of chronic stress.

The act of engaging in a therapeutic ritual, like crystal healing, may help individuals cope with stress-induced health conditions. Believing that you are taking charge of your health could increase feelings of control that in turn reduces stress. As the stress hormones decline in our body, you may feel more relaxed, sleep more soundly, digest your food better and recover from pain. In effect, the balance is restored between the gas pedal and the brake pedal and you can cope with challenges more effectively. It is no coincidence that many alternative healing practices are purported to help treat stress-induced health conditions, like chronic pain and irritable bowel syndrome (IBS). Likewise, support for the placebo effect is most pronounced for similar health conditions. But crystal healing, and there-fore the placebo effect, only exhibit any effectiveness with these kinds of stress-induced conditions. Medical conditions that are caused by bacteria or viruses or cancer are not going to be directly helped by placebo or alter-native healing treatments.

3. Patient's response to interacting with a practitioner One of the strongest contributors to the placebo effect is the patient-doctor interaction. Studies on the placebo effect show that patients' health improves the most when they interact with doctors who respond to them in an empathic and warm manner. This includes doctors who encourage patients to describe their symptoms and to elaborate on their relationships and lifestyles. Patients more likely showed improvements in their symptoms when doctors conveyed confidence and positive expectations for the treatment and listened to their patients' concerns.

❝People may engage in alternative healing practices as a way to compensate for what is missing in conventional medical settings.❞

Medical visits in the formal doctor's office setting are rarely as in-depth and empathetic as they should be. Doctor-patient consults are often rushed and so the rapport-building that can improve treatment outcomes is often neglected. People may engage in alternative healing practices as a way to compensate for what is missing in conventional medical settings. When asked why they practice alternative medicine, most people claim that they did so for the interpersonal connection, the sense of empowerment, and the confidence in the treatments. Being part of the crystal-healing community may help individuals find others with similar desires and beliefs, providing an opportunity to build meaningful social relationships. Scientific research has shown that positive social relationships are beneficial for health. For example, cancer patients assigned to support groups have better survival rates compared to patients given conventional treatment minus the support group. Those who feel less lonely are less likely to come down with cold or flu symptoms and less likely to die early compared to those who feel more socially isolated. Interpersonal connections made through the practice of crystal healing could therefore be a major factor in improving mental and physical health.

Despite continued support for science and medicine, many people hold pseudoscientific and metaphysical beliefs about crystals and other alternative healing practices. Some proponents believe that crystals can tackle anything from common maladies to personal worries, and most problematically, they also believe that they can cure serious medical conditions like cancer. Personal experience and anecdotal evidence continue to support the methods and practices of crystal healing. When you are faced with compelling patient testimonials, it is important to remember that it is the belief in crystal healing, rather than the physical properties of crystals themselves, that is driving causal illusions and purported positive outcomes.

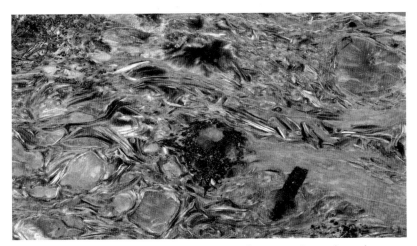

It's easy to see why charoite is wildly popular among collectors, jewelers, and crystal healers alike. This now hard-to-obtain Russian gemstone fetches high prices today.

Crystal healing in no way can substitute for evidenced-based treatments for serious medical conditions like cancer. Instead, it may serve as a useful tool for self-improvement and connection with others. Yet, self-improvement is not dependent on crystals, but instead derives from changes in our physical health, thought patterns, and emotions. Crystals do not impart energy, but instead they may serve as reminders to be more thoughtful and to be more empathic toward others. Certain practices of crystal healing may also help people slow down, breathe deeper and feel more in control, thus reducing the negative consequences of stress. In sum, the crystals themselves are unnecessary for improving our lives. There are other evidence-based and more cost-effective practices that are much more effective tools to positively impact our health and well-being. These include activities like mindful breathing, connecting with friends and family, and community engagement, such as volunteering. Crystal-healing practices may overlap with some of these activities, and therefore with some of their positive effects, but this is a classic case of correlation, not causation.

Among the most common minerals on the planet, calcite is an inexpensive collectible.

Feldspars are the most common group of minerals on Earth, found in the majority of rocks, yet are often selectively ignored by crystal healing authors. While the prettiest feldspars—moonstone and labradorite, for example—are considered crystal-healing essentials, the common feldspars are rarely given any attention, due to their dull colors and often less-than-spectacular crystals.

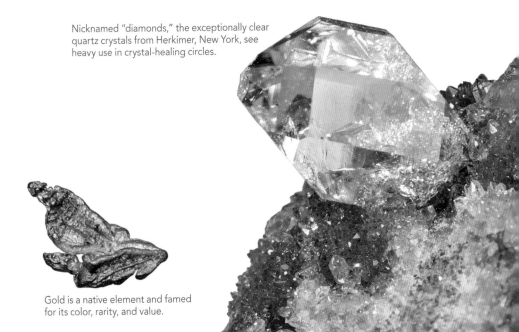

Nicknamed "diamonds," the exceptionally clear quartz crystals from Herkimer, New York, see heavy use in crystal-healing circles.

Gold is a native element and famed for its color, rarity, and value.

SCIENTIFIC SPOTLIGHT: SAGAN'S DRAGON

Pseudoscience's proclamations of invisible energies and intangible connections to the universe have long troubled real scientists. Such baseless claims mislead the public about real scientific breakthroughs and diminish the importance of proper scientific work in favor of quick fixes or non-existent magic bullets. And worse, because pseudoscience is rooted in a reality of its own invention, devotees can easily refute information contrary to their beliefs by merely fabricating yet another untestable claim on the spot. By confidently asserting their beliefs while simultaneously putting up road blocks between them and any skeptical inquiries, practitioners of pseudoscience pretend to have legitimacy and convince themselves and their followers that their and poorly defended "science" is accurate. Our natural desire to confirm our own beliefs is exploited by pseudoscientists in order to sell us their goods and services, while real scientific discovery—which is perhaps "too slow" for the tastes of today's internet self-help culture—goes underappreciated.

The dangers of pseudoscience and the disservice it does to those misled into believing it was not lost on noted astronomer and science communicator Carl Sagan. Sagan spent much of his life working tirelessly to open the public's eyes to the wonders and importance of real science. In his 1996 book, *The Demon-Haunted World: Science as a Candle in the Dark*, he proposes a hypothetical situation which highlights both the way pseudoscientists counteract their skeptics and the outlandish nature of pseudoscientific beliefs in general.

Sagan conjures a scenario in which he claims to you, the reader, that there is a fire-breathing dragon that lives in his garage. Incredulous, you would ask to see it, and upon opening the garage door, you indeed see no dragon at all. Met with your skepticism, he explains that the dragon is invisible, but assures you that it exists.

You may suggest spreading something onto the floor that would make the dragon's footprints evident, to which he replies that the dragon actually floats on air and doesn't touch the ground. Then you may suggest using an infrared thermometer to locate and measure its invisible fire breath. He counters by saying that the fire actually doesn't possess any heat. Frustrated, you suggest painting the dragon to make it easier to spot. He responds this time by telling you that it has no physical body. He continues to counter every test you suggest by concocting a reason or explanation as to why it wouldn't work, yet insists the dragon does indeed exist in his garage. He offers no other evidence beyond his own belief in the dragon.

The question then becomes, "What's the difference between an invisible, heatless dragon with no body and no dragon at all?" Indeed, the same can be asked of crystal healing—why should anyone believe in invisible, unmeasurable, untestable energy emanating from rocks and minerals, and why would it heal us if it did? In Sagan's thought experiment, you cannot prove that his dragon doesn't exist, despite all of the evidence to the contrary. However, the inability to say conclusively that it doesn't exist does not mean that you should then say it does exist.

Yet this is the kind of tenuous assertion that pseudoscientists make every day. The same situation happens with crystal healing—the fact that we cannot measure or test these "healing energies" should be evidence that they don't exist, yet is instead taken by crystal healers to mean that they have not yet been disproved.

With Sagan's dragon experiment, you may be left wondering about the person who believed in the dragon, something that was seemingly nonexistent. But, as Sagan continues to propose in his book, what if he weren't the only one to believe in the dragon? What if many people assured you, the skeptic, that they, too, knew that such dragons existed? If they could prove that they had all seen footprints appear or felt fiery dragon breath first-hand, or show you directly, then that could be evidence that you may be wrong. But if such evidence—a dragon's footprint or a burn from its breath—appeared while no skeptic was present and could not be replicated, then in fact no real evidence has been proposed at all. And in the face of a lack of grand evidence for such a grand claim, the only sensible approach is the scientific one: to reject the hypothesis (at least for now), be open to future findings, and give some serious thought to why so many seemingly normal people would share the same misguided belief.

In addition to his work as a science communicator, Carl Sagan had deep ties to NASA. When the Voyager 1 space probe left our solar system in 1990, Sagan was instrumental in convincing NASA to turn its camera back toward Earth to capture our planet's "portrait" from more than 3.5 billion miles away. The result was the famous "Pale Blue Dot" photo, shown here; Earth is the tiny bright speck near the center of the photo.

A QUICK GUIDE TO COMMON CRYSTALS

Many of crystal healing's "most powerful crystals" are those of common and/or commonly available minerals. This makes sense; if only the rarest and most valuable minerals were considered the best for healing, few people would be able to participate. But quite a few of the "crystals" frequently used in metaphysics are not crystals at all, but are rocks or glass. In other cases, commonly used materials may be dyed, altered, or altogether fake. As always, a little research can go a long way, and in this chapter we'll discuss some of the most popular materials used in crystal healing and what everyone should know about them.

"Many of the 'best healing stones' are simply the ones most affordable and easy to obtain. This is no coincidence; it's hard to sell a lifestyle if its best tools are also rare and expensive."

THE HEALER'S TOOLBOX

When it comes to determining which minerals' crystals are "best" for healing, crystal healing books will often have differing lists, but there is a general consensus around the top 20 or 30 "most powerful" species for the task. Like everything else in crystal healing, though, these lists are rife with fundamental misunderstandings of basic geology and mineralogy, and vague yet inaccurate (but fancifully worded) accounts of where and how minerals form. This may inspire feelings of reverence and wonder in the reader but ultimately does them the disservice of misinforming and misleading them. Once again, believers in New Age metaphysics disregard well-understood science in favor of invented mysteries, all in the name of selling books, and overpriced crystals and augmenting their lifestyles.

A common theme among crystal healing authors when discussing how minerals form is an extremely generic description such as they are "born deep in the earth," and are "unimaginably ancient." While it is true that some minerals are extremely old—billions of years, in some cases—and some, such as diamonds, did form deep within the earth, these are far from being universal facts. Many form at or near Earth's surface, and some crystallize rapidly, sometimes within a matter of days or weeks. Gypsum—often regarded as a particularly powerful mineral—actually does both. You, too, will easily see that the facts defy the claims of crystal healers if you take the time to research minerals for yourself.

While you can easily find elements of hypocrisy and pretense throughout all aspects of crystal healing, there is perhaps no greater example than when the mining of minerals is considered. Most metaphysical practitioners pride themselves on being in-tune with nature and respectful of our earth, hence their reverence for crystals. However, crystal specimens are almost always obtained through some form of mining, whether large-scale industrial strip mines or small rural dig sites, often in sensitive environments like jungles or tundras. Mining is inherently invasive, and many of the most affordable, commonly available crystals are sourced en masse from Africa, South America, and Asia, where mining laws, safety standards, and environmental considerations are often lacking and/or poorly enforced. Sometimes outright illegal and unsafe mining practices are employed. Amber, for example, is widely popular in crystal healing, but sources in the Dominican Republic obtain specimens from dangerously hand-dug mines prone to collapse. In Russia and Ukraine, mining interests are often controlled by organized crime, and amber is collected illegally by force. And around the world, countless fake amber specimens abound in the marketplace to deceive collectors.

Today, many crystal healers are urging their readers to "make sure their crystals are ethically sourced," but what is "ethical?" Most minerals and their crystals are not found lying loose on the earth's surface, and the vast majority must be mined to satiate the market. Clearly those using crystals in an attempt to live more holistically should know about these questionable aspects of mining and how to avoid such crystal sources. Nonetheless, this information is often lost, glossed over, or willfully disregarded in pursuit of getting affordable specimens to the market.

Altered crystals are another concern. The industry promotes the concept that the biggest, finest, and clearest crystals will be the most effective in healing. Not coincidentally, these are also the most expensive crystal specimens. For example, quartz is extremely common and its crystals are far from rare, but the emphasis on large, perfect specimens is great enough that specimens are frequently "fixed," using saws and polishers. Similarly, common minerals with drab coloration may be dyed or otherwise altered to make them more vibrant and salable. Ironically, crystal healing, which places so much emphasis on the natural properties of minerals, is rife with altered or treated crystals, but often the authors recommending such crystals aren't even aware of these common practices.

The preceding issues and misconceptions aside, many crystal-healing practitioners consider crystals to be "alive," and claim they should be viewed as our "partners" in healing. Language regarding the "birth" of the crystals and "earning their respect" is commonplace as well. They take well-understood phenomena—namely the seemingly mystical geometry of crystals—and attribute them to constructions of divine forces that govern the universe. These claims exist in their own version of reality—one in which the definition of "life" is extended to inorganic substances, apparently—and as with all pseudoscience, devotees use the lack of solid evidence against their claims as evidence for them. They live in a reality of their choosing.

These quartz crystals may look attractive, but are all shaped by hand; they were cut and polished from larger, cruder masses to create "ideal" quartz crystals for healing.

Or do they? More and more, hidden on the back pages of crystal-healing books or in small print at the bottom of a web page, you can find disclaimers in the writings of crystal-healing gurus, stating that "crystal healing is no substitute for consulting a healthcare professional" or to "see a doctor before beginning any alternative medicine treatments." Some proponents may say that these disclaimers are included to protect the writers from skeptics (and lawsuits), but the possibility also exists that they include it because the authors don't actually believe what they claim. Perhaps disclaimers like these show that the crystals they sell, the advertising revenue their websites earn, and the books they peddle are all elements of a successful business. Or perhaps they don't want a serious illness or death on their conscience when their spurious claims lead someone to cease treatments for their curable cancer, for example, in favor of crystals that do nothing to combat the disease. In all cases, when a lifestyle is being purported as a cure-all or as an answer to society's woes, we must be ever-vigilant and skeptical, asking ourselves, "What are they gaining from this?"

THE LIST

Many of the "best healing stones" are simply the ones most common on the market; they are often more affordable and easy to obtain. This is no coincidence; it's hard to sell a lifestyle if its best tools are also rare and expensive. So minerals like quartz and its many varieties, found all over the world and frequently bought and sold by the barrel-full, are used for all kinds of metaphysical remedies. Some of the choices for these "best stones" are seemingly arbitrary or are based on a misunderstanding of what the minerals actually are. Jade, for example, is a term that refers to the translucent green gem form of a couple different minerals. "Jade" therefore is a description, not a particular mineral, and yet "jade" is often emphasized for its healing properties. Which jade? A crystal healer may tell you that it doesn't matter, but if different types of "jade" can be composed of different minerals, then it surely should matter which one you use, right? (This is like treating all painkillers the same; aspirin and ibuprofen both reduce pain, but they are hardly the same drug.) Or perhaps crystal healers who instruct you to use "jade" are simply so misinformed that they don't understand the difference themselves.

Here we will look at a selection of "powerful" rocks and minerals and investigate healers' claims versus reality. The minerals on the following list are considered by many crystal healers as some of the "most essential" crystals to add to your collection. All are said to have important healing properties, but the most lauded and widely cited crystals are listed first.

BUYER BEWARE

When browsing mineral specimens for sale, whether your intent is to use them for healing or just to collect them, there are a few key alterations, misrepresentations, and inaccuracies common in the marketplace (yet rarely disclosed to the buyer) that you should be aware of.

Some specimens on the market may not be naturally shaped, but have been cut, carved, and polished to have a more idealized crystal shape. This is done to enhance the appearance of crystals that may not have formed ideally, but is also done with rocks to give them a more appealing crystal-like shape, despite the fact that rocks do not form such shapes.

Some crystals or crystal clusters may have been repaired or stabilized in order to make them more salable. When large crystals are extracted, there's a good chance that their edges or points could be damaged in the process. Careful cutting and polishing can "fix" these breaks to the untrained eye. Similarly, some crystal clusters and other mineral formations may be glued together or injected with resin to stabilize them and prevent them from crumbling.

Many common rocks and minerals are colored to make them more exciting and salable. Dying, heating, and irradiating are common methods used to color crystals. Researching the typical appearance for such minerals is a key first step in learning to spot altered color.

Many minerals share similar appearances, and common specimens may be misrepresented as a rarer and more valuable species. Similarly,

Smoky quartz is favored among crystal healers and mineral collectors alike, particularly specimens containing "phantoms"—colored zones inside a crystal that developed when levels of impurities varied during crystal growth.

a dyed stone may appear similar to something rarer and be sold as such. Sometimes this is done unwittingly, but other times is done on purpose. Research will be your best defense.

The majority of rocks and minerals sold at a typical rock shop, tourist shop, or alternative healing store are not rare. Regardless of how they may be marketed, the average specimen in these places can be acquired in bulk. While knowledgeable shops will certainly have valuable rarities, the polished stones and small crystals in little padded boxes that line store shelves are common and should be priced accordingly.

Many common rocks and minerals are given creative names—sometimes trademarked—to enhance their story and salability. If you can't find the name of a mineral in a scientific source, then it has likely been invented to sell something common.

All buyers should be aware of the unethical mining practices used to obtain certain minerals, including the use of child labor, ecologically destructive mining, and armed conflict in securing crystals. Most minerals have few issues in this regard, but many of the most popular healing stones, such as rose quartz and diamonds, should be thoroughly researched to avoid purchases that could support these problematic practices.

Under the right conditions, goethite can develop a rainbow-like surface sheen, turning this otherwise dull iron ore into something spectacular.

These finely developed, sharp blades of crystallized gypsum formed in mineral-rich water at an old mine site, tinted green by copper.

Selenite (also called satin spar)

What they claim Selenite is often called a "master" crystal, noted for its particularly powerful energy, especially when used on the upper chakras around the head. As such, it is said to clear your mind and help with mental problems. Selenite is also said to be the best mineral for connecting with "ethereal" planes and spirits, and is frequently claimed to align the spine and protect against epilepsy.

What it actually is Selenite is the name given to crystallized forms of the mineral gypsum; therefore, selenite is gypsum. Gypsum, the most common sulfur-bearing mineral and among the most abundant minerals on Earth, forms in a number of ways but most often as a result of precipitation, when mineral-rich water evaporates and leaves minerals behind, in low-temperature and low-pressure conditions at or near the earth's surface. Typically gypsum forms massively, which means that it appears in large masses rather than fine crystals. Gypsum beds, or layers within rocks, can be enormous and are composed of fine crystalline grains of gypsum compressed together, often with numerous impurities and other evaporation-related minerals mixed in. Fine examples of selenite can appear as transparent blades or wedge-shaped crystals, often with a distinctly fibrous texture. Given enough time in just the right conditions, crystals can grow to many feet in size, though the vast majority are under 12 inches in length.

OPPOSITE: Enormous beds of white gypsum are mined for the construction industry.

What healers should know Selenite is merely the fibrous, finely crystallized form of gypsum, formed under ideal crystal-forming conditions. Chemically, there is no difference between selenite and typical gypsum. (Incidentally, gypsum is also the main component in many forms of drywall, which few, if any people, would consider a "healing" material.)

Any proclaimed differences between the effectiveness of selenite and gypsum are therefore as arbitrary as the claims themselves. Most importantly for healers, while some examples certainly are thousands or millions of years old, a great number of the finely crystallized selenite specimens on the market are post-mining formations, which means that they grew in wet or flooded mining tunnels or holding ponds after mining had ceased. Crystals can also grow very quickly, given the right conditions. Therefore, some specimens of this "ancient master mineral" are merely a few years— or just a few weeks—old.

NATURAL APPEARANCE

WHAT'S COMMONLY SOLD

VS

A large, clear selenite crystal from a sedimentary environment

Selenite "towers" are often sold as natural formations for metaphysical uses, but have actually been shaped by hand.

Common massive gypsum with no visible crystal shape

Fibrous selenite pieces like this are known as satin spar, due to their satiny sheen.

Feldspars are most prevalent in igneous rocks like granite, where they often appear as light-colored blocky masses.

Moonstone

What they claim Moonstone has long been regarded as a sacred stone, claimed to facilitate visions of the future and interact with one's soul. It is also considered a "feminine" stone, said to aid in one's love life, emotional balance, and in easing menstrual discomfort.

What it actually is Moonstone is a variety of feldspar, the most abundant group of minerals in the earth's crust. Consisting of several separate species, the name "feldspar" refers to minerals with a similar chemistry and crystal structure, including orthoclase, microcline, and labradorite. Most feldspars are prevalent as constituent grains of rocks, especially igneous rocks like granite. "Moonstone" is not a mineral itself but is a descriptive term for a whitish or translucent feldspar, usually microcline or albite, that has grown in such a way as to include micro-layers of another mineral (often another feldspar) between the thicker layers of the primary feldspar. This results in incoming light bouncing between these differently crystallized layers, which produces a bluish or whitish schiller, a "flashy" optical phenomenon when the crystal is rotated under bright light. The schiller is seen just below the surface of the stone, giving it the appearance of a crystal glowing from within, which has captured the imagination of people for centuries.

What healers should know As with many mineral varieties, the name "moonstone" doesn't refer to any particular mineral, but rather a specific appearance that some feldspars that formed in specific conditions can exhibit. Therefore, if healers insist that each mineral has its own set of healing properties, so too should moonstones of different feldspar composition. There is also the question of why moonstone (and other attractive feldspars) is more emphasized than typical feldspar examples. Feldspars are extremely abundant, present in a majority of igneous rocks, and easily found virtually anywhere in the world, yet are largely ignored by crystal healers. Does their abundance not mean that the universe deemed feldspar most important? Or perhaps the common (and usually quite plain) feldspars simply are not marketable enough; selling the most common mineral on the planet might prove a tough sell, even for purveyors of healing crystals.

NATURAL APPEARANCE

VS

WHAT'S COMMONLY SOLD

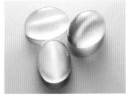

A fine rough specimen of bright moonstone.

Though most people can easily spot the difference, fiber-optic glass is sometimes polished and sold as inexpensive moonstone because of its somewhat similar play of light.

A typical polished piece of moonstone showing a strong white internal schiller at this lighting angle

A specimen of rough moonstone showing a bluish-white internal schiller

The same piece of moonstone as above, but the lighting has changed angle and the schiller is far less apparent

The same piece of moonstone as above, but rotated in such a way that the schiller is no longer visible

A vividly colored specimen of rough aventurine.

Aventurine

What they claim Aventurine is usually said to be a stone that promotes decisiveness and leadership qualities, and one that inspires optimism. It is also said to bring good luck to its holder while promoting empathy, alleviating neuroses, and enhancing creativity. In general, aventurine seems to be an all-purpose stone.

What it actually is Aventurine is a variety of quartz that contains countless tiny inclusions of a green mica mineral, usually fuchsite (the green variety of muscovite). The micas are a group of closely related minerals with crystals that are shaped as flat flakes or plates, often formed together in stacks. Micas are most common as constituent minerals in rocks. Aventurine forms massively and includes the tiny, shiny flakes of mica dispersed among crystalline grains of quartz, giving the mass of quartz a fairly even greenish coloration and often a "glittery" sheen. Both quartz and mica minerals are very common and frequently occur together in various rocks, but not usually with such attractively colored results. Aventurine does not form under conditions that produce fine quartz crystals, and therefore any specimen of aventurine appearing as a well-formed crystal instead of an irregular mass has been shaped by hand. The vast majority of specimens on the market originated in India and have been polished and shaped by Asian and Russian artisans.

What healers should know Most mica minerals garner little attention from crystal healers, as most are fairly unremarkable in appearance, usually only seen as small flakes in rocks such as granite. Why the mica seen in aventurine is something more "powerful" therefore must be a matter of its attractiveness. But aventurine is two separate substances—quartz and mica—and can be considered a rock, so why should it have its own distinct set of healing traits? And if well-developed crystals are most useful for healing, why should aventurine, which is not found as free-standing crystals, be regarded as so important? Aventurine specimens aren't just polished and carved to make them more salable; some pale aventurine is also dyed deeper shades of green to make it more attractive, or may be dyed to different colors entirely. Blue aventurine does exist naturally, but red, purple, and other more exotic colors are almost certainly altered (i.e., unnatural), even if not advertised as such (because the shop-keepers themselves may not know).

NATURAL APPEARANCE

WHAT'S COMMONLY SOLD

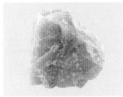

VS

A common natural specimen of somewhat pale aventurine

A tumble-polished and color-enhanced (dyed) pebble of aventurine

Tumble-polished pieces of aventurine are common and affordable in most every rock shop.

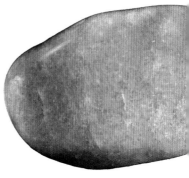

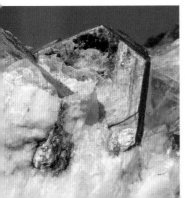

This finely crystallized muscovite shows the hexagonal shape most micas take when they have enough space during formation.

Naturally yellow quartz, called citrine, is very rare; far rarer than most crystal healers would have you believe.

Citrine

What they claim Citrine is usually said to promote clarity of thought and creativity and can cleanse the chakras. But more often it is associated with abundance and said to invite wealth into your life when placed in particular positions around your home. It is also claimed that it helps with eye problems and infections and can "reverse" degenerative diseases.

What it actually is Citrine is the name given to the naturally yellowish variety of quartz. The source of the color is still not known for certain; the most likely cause is irradiated impurities of aluminum, very similar to the cause of the gray-black coloration in smoky quartz. In fact, many natural citrine crystals have zones of smoky coloration within them. But natural citrine is far rarer than smoky quartz, and in fact is rarer than most other colored varieties of quartz. Its crystals are usually small and many are very pale in color, but other specimens may be so dark as to appear brown (which is a good example of how subjective color variations can be).

What healers should know Natural citrine is exceedingly rare. The "citrine" crystal points found in rock shops, visitor centers, and toy stores that are frequently priced around one dollar or less are actually heat-treated amethyst. Amethyst, or purple quartz, will change color to a deep honey-yellow when heated sufficiently, and due to the incredible abundance of low-grade amethyst, particularly from Brazil, enormous amounts are heated to make them more salable as "citrine." Real citrine is often more lemon-yellow, brown-yellow or greenish-yellow in color, while the heat-treated amethyst "citrine" usually has a fairly uniform burnt-yellow or orange-yellow color. Heated citrine crystal points also usually have jagged, broken bases, from where they were separated from the rest of the mass in which they originated, while real citrine crystals may have better defined crystal faces all along their length. But the easiest way to tell a heated citrine from a real one? Aside from doing proper research into localities that produce true citrine, the price in a mineral shop will be a telling clue: heated amethyst is very abundant and affordable, while even a small, but nice, crystal, of real citrine could fetch a hundred dollars or more. Most of the time, however, shop keepers don't even know that what they're selling as "citrine" is actually altered amethyst; suppliers don't always volunteer that info.

NATURAL APPEARANCE   *WHAT'S COMMONLY SOLD*

VS

A very rare natural citrine crystal exhibiting a yellow-brown coloration

A typical rock shop "citrine," showing the telltale amber coloration of heat-treated amethyst and ragged base

Crusts of amethyst crystals from Brazil, before treating

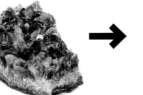

A crust of amethyst crystals after heat treating, sold as "citrine"

A cut citrine gem of highly desirable canary yellow color

A cut and polished section of nephrite, showing the typical mottled and variegated coloration

Jade

What they claim Jade is said to symbolize serenity, promoting peace and wisdom. It also has a strong association with dreams, general well-being, and nature, said to draw upon Earth's life force if you wear it outside. Emphasis is also placed on other color variations of jade, each said to provide different effects.

What it actually is The term "jade" is more of a description than it is a particular mineral. "Jade" refers to the solid gem form of two distinct minerals: jadeite (a mineral of the pyroxene group) and actinolite (a mineral of the amphibole group). Both minerals typically form as grains within rocks or as hard, fibrous elongated crystals, but under the right conditions they will form massively (without a crystal shape) in the semi-translucent aggregates we call jade. Jadeite is the rarer of the two, often exhibiting lighter coloration, while actinolite jade, also called nephrite, is more available and comes in deeper, richer green hues. As both types of jade are frequently impure and contain other minerals, "jade" is generally considered to be a metamorphic rock, and therefore forms no crystals.

What healers should know The fact that jade refers to two distinct minerals means that the healing importance placed on "jade" is completely arbitrary, as pyroxene and amphibole minerals individually are usually said to be good for different aspects of healing. The longtime popularity enjoyed by jade also means that other green minerals are frequently misidentified as jade, both accidentally by those hopeful to find it and intentionally by those looking to make a less desirable mineral more salable. Various forms of green quartz, including aventurine, are the most frequent lookalikes, but they are harder than both forms of jade and will take a better polish. Sometimes quartz or other minerals are dyed to look like jade, but will often have too uniform a color; real jade often has mottled coloration, with variations of greens, usually with browns, tans, yellows, and black spots. Minerals of the serpentine group, also often deep green, have also been marketed as jade due to their many visual similarities, but in general they are much softer than true jades. Nephrite jade can also have outer surfaces that appear waxy, or softly lustrous, while most fake jade of harder composition (e.g., dyed quartz) will often be polished to a bright, glassy shine.

NATURAL APPEARANCE

WHAT'S COMMONLY SOLD

VS

This rounded, beach-worn specimen of nephrite is a prized discovery along the Pacific Coast of North America.

Other green minerals, such as this richly colored aventurine, are often sold as "jade."

A specimen of dark greenish serpentine, the type often confused for jade

A rough specimen of jadeite, showing the typical pale green coloration, in contrast to nephrite's richer hues

"Tibetan quartz," often placed in a position of great importance by crystal healers, is merely typical quartz, and is rarely, if ever, from Tibet.

Quartz

What they claim Quartz is often said to be the most powerful healing stone, capable of nearly any metaphysical task. With particularly large crystals often used to "charge" smaller ones, quartz is also said to be linked directly to the spirit realm and therefore can be used to cleanse our souls and for "interdimensional travel."

What it actually is Quartz is the single most common mineral in the earth's crust; it is present as a constituent in many kinds of rocks, particularly granite. Composed of silicon dioxide, or silica, it is very hard and brittle, has a glassy luster, and is usually translucent and colorless, unless tinted by impurities or other elements in its structure. Smoky quartz, for example is also popular in crystal healing and is colored by naturally irradiated aluminum within the crystal structure. Quartz crystals can take several shapes depending on the environment in which they formed, but they are most commonly seen as hexagonal barrel-shaped crystals that are longer than they are wide, and which terminate, or end, in a pointed or chisel-shaped tip. While crystals are very common, particularly as coatings or crusts of countless tiny crystals called druse, it is even more abundant in massive (non-finely crystallized) forms: enormous masses of quartz form deep in granite bodies, thick veins of milky quartz crosscut metamorphic rocks, and most of the world's rivers are home to countless quartz pebbles. Because quartz is inert and extremely weather resistant, it takes millennia for it to undergo any change, but it eventually ends up as grains of sand. As a result, quartz is virtually always underfoot in one form or another.

What healers should know Quartz is ubiquitous in crystal healing; large, very finely formed crystals are common, and their sharp shapes and clarity do well to inspire both natural wonder and belief in their connection to the spirits. But espousing the supposed healing benefits of quartz is little more than clever marketing, turning the most abundant and commonly available mineral into something "essential" for believers. Small but well-developed quartz crystals are bought and sold by the barrel-full and distributed by the truckload, and they are sourced from all over the world. Rock shops usually sell small specimens for a few dollars or less, as is appropriate given their abundance, while crystal-healing shops and websites frequently charge many times the common price. But quartz can easily be obtained without the need to spend any money at all; a trip to a river or beach will usually yield more pebbles of quartz than you can carry.

NATURAL APPEARANCE

WHAT'S COMMONLY SOLD

 **VS**

Natural quartz crystals are abundant. Their sides are rarely all exactly the same width or length, and their tips are often chisel-shaped rather than sharp points.

Large crystal points of ideal shape, such as this one, have been cut from irregular masses of quartz and polished to appear as perfectly formed as possible.

A finely formed, dark smoky quartz crystal

Tumble-polished rose quartz is cheap and widely available.

Rose Quartz

What they claim Rose quartz has always been associated with love and is said to invite love into one's life and help maintain current relationships. By extension, it is also claimed to be beneficial for heart health and to soothe tension and relieve stress. It also sees heavy use in healing, including for treating lungs, fertility problems, and vertigo.

What it actually is As its name implies, rose quartz is the naturally pink variety of quartz. There are actually two kinds of pink quartz. The more common of the two forms in the core of granite pegmatites, which are the deepest and most coarsely crystallized portions of granite formations. Here, it forms as huge masses without any visible crystal shapes. Fine examples are deep, vivid shades of reddish-pink, but the majority is fairly pale and can be nearly opaque. In this setting, the pink coloration is caused by countless microscopic fibers of a pink boron-bearing mineral, which lends the quartz its color. This kind of rose quartz is very common in the marketplace.

The other variety of rose quartz also forms in pegmatites, but in open pockets that developed very late in the solidification of a pegmatite formation. This kind of rose quartz develops as fine crystals, often in elaborate clusters, and can show exceptional clarity and a soft pink coloration. In these specimens, the pink coloration is caused by irradiated impurities of aluminum and phosphorus. Due to the very different origins of its rosy hues, mineralogists call this variety "pink quartz" to distinguish it from the more common rose quartz. This kind of rose quartz is far rarer and specimens are much more valuable; it is also far less common in crystal healing literature. Crystal healers using the term "rose quartz" are virtually always referring to the common massive form.

What healers should know The common variety of rose quartz is usually found in enormous masses weighing several tons that must be broken up for sale. Two of the world's leading sources for rose quartz are Brazil and Madagascar; both countries have had spotty histories of questionable mining practices, poorly regulated safety guidelines for miners, and mining being done in ecologically sensitive areas. In Madagascar, illegal mines employ entire families—including children—to extract rose quartz from deep, narrow, hand-dug tunnels that are prone to collapse. This dangerous practice supplies the alternative healing industry with most of its rose quartz while paying the workers only a few cents per kilogram of quartz hauled out of the mines by hand. These social realities and environmental consequences should be considered by everyone who buys minerals—not just crystal healers.

NATURAL APPEARANCE   WHAT'S COMMONLY SOLD

 VS

Typical rough mass of the common type of rose quartz   *Common rose quartz does not form as crystal points, but it is frequently carved into a crystal-like shape.*

A specimen of the far rarer "pink quartz," exhibiting stubby but well-formed crystals

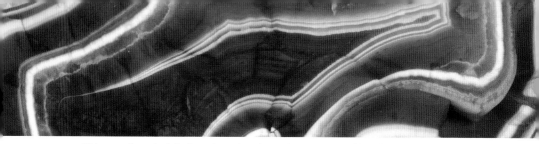

This agate from the Lake Superior region shows the quintessential banded pattern and vivid coloration prized by collectors.

Agate

What they claim Agates are said to promote inner stability and harmonize positive and negative forces. They are also claimed to increase concentration and make the user a more truthful person, while creating a feeling of safety in the holder.

What it actually is Agates are one of the more enigmatic mineral formations commonly available in shops. They are a variety of quartz—namely a microcrystalline variety called chalcedony—that develops as nodules containing onion-like layers. Each layer contains the smaller layers within it, concentrically arranged around a common center. This banding, as it's known, can have beautiful alternating colors and differing textures and levels of translucence. But despite their popularity as a decorative stone for several thousand years, we still don't know how they formed. While there are many hypotheses, some more convincing than others, none has been able to sufficiently explain agates and all the various structures and patterns we see within them. Agates and their curious band-within-a-band arrangement of chalcedony layers form within the gas bubbles in volcanic rocks, namely basalt, but also rhyolite, but not in all formations of these rocks. Clearly their formation was dependent on very specific conditions, but so far there is no consensus on which conditions are most conducive to agate formation.

What healers should know Agates come from all over the world; they have been found on every continent, and can exhibit nearly any color and all manner of exotic patterns. But poorly banded specimens with unexciting gray and brown coloration are most common, particularly from Brazil. Brazil's agates, some of the youngest on the planet, are commercially mined, with hundreds of thousands of specimens produced every year. Yet only the most colorful are salable on their own—the less exciting examples don't often garner much attention. In order to make them more salable, these dull specimens are altered in different ways to increase their appeal, including heating them to bring out iron-rich red colorations and dying them blue, green, pink, and purple. These colorations may be more exciting on a store shelf, but all-too-often these are not advertised as altered stones. And it isn't just Brazilian agates that are treated; agates from Uruguay, Madagascar, and Mexico are also often altered for salability, some more subtly than others.

NATURAL APPEARANCE

A finely banded agate specimen, naturally rounded by wind and waves

VS

WHAT'S COMMONLY SOLD

Overly vivid agates in shades of purple, blue, pink, and green are dyed

These orange agate nodules still embedded in their host rock are whole; no banded pattern can be seen until they are cut open.

Brazilian agates with typical dull coloration are frequently dyed vivid colors to make them more salable.

These tiny green crystals are uvarovite, one of the most sought-after garnets.

Garnet

What they claim Garnets are said to be strong healing stones, promoting regeneration and physical mending, and are symbols of health in and of themselves. They are also said to enhance creativity and aid in building relationships, and are claimed to enhance one's sex life. Some garnets are even said to treat cancer, leukemia, and strokes, all of which are dangerously misguided beliefs.

What it actually is The term "garnet" is the name for a group of minerals and therefore does not refer to any individual mineral. When "garnet" is written, usually the author is referring to almandine, pyrope, spessartine, andradite, grossular, or uvarovite—these are considered the "traditional" garnets, though the garnet group contains more than a dozen minerals in all. Garnet minerals are related chemically and all share a similar crystal structure, often appearing as faceted ball-like crystals embedded in rocks, particularly in granite or metamorphic rocks like gneiss or schist. Many are reddish, but more are brown, yellowish, or black, while some rarer species, such as uvarovite, are green. In general, garnets are all hard and glassy, and have been used as gemstones since antiquity. Almandine is the most common, and virtually any deep reddish garnet can be assumed to be almandine until proven otherwise.

What healers should know More often than not, crystal-healing books simply cite "garnet" as the crystal to use. But which garnet? There are so many, and all contain different elements and different physical properties, so surely they can't all be good for the same healing tasks, especially when quartz and its colored varieties—all of which are virtually chemically identical—are said to be good for vastly different healing purposes. In older mineralogical texts, authors rarely differentiated the common garnets due to the difficulty of doing so and because differentiating between them wasn't necessary for general discussions of rock components. As the crystal-healing community began developing their lists and techniques, it was likely from these less technical books that they acquired their understanding of minerals. But as crystal-healing books continue to include only "garnet," it should signal to the reader that the author has insufficient mineralogical knowledge to pretend to have any authority on the subject.

NATURAL APPEARANCE

VS

WHAT'S COMMONLY SOLD

A natural crystal of almandine, one of the most common garnets

Common, crudely formed garnets are often "cleaned up" by carefully polishing the faces, resulting in a nice, but altered specimen.

A gemmy grossular still attached to its host rock

A particularly large and well-formed cluster of andradite crystals

Clusters of vivid amethyst are popular and readily available

Amethyst

What they claim Amethyst is often called a stone of manifestation, making it easier to make decisions and to set and attain your goals. It is also a "shielding" mineral, providing protection against negative energy, and aids in ridding oneself of vices, such as alcoholism. It has been used since antiquity to guard against drunkenness.

What it actually is Amethyst is the name for the purple variety of quartz, and is one of the more common colorations of quartz found around the world. Its purple color, caused by irradiated iron impurities contained within the structure of the quartz, is usually somewhat uneven within a crystal, with the most intense coloration being a few millimeters beneath the crystal's surface. Amethyst crystals are rarely very long, and most are small, coarse crystal points clustered together in layers or crusts called druse. Some of the most famous examples of amethyst are the stunning amethyst "cathedrals" from Rio Grande do Sul, Brazil. (They are very popular on the market as well.) These geodes (hollow mineral formations) are nicknamed "cathedrals" because of their tall domed interiors. They formed within basalt lava flows, but how they formed is still a matter of some debate. They likely resulted from fluids, possibly steam, rising upward through the lava flow as it reacted with the underlying rock. These and other Brazilian specimens often have rich coloration and dominate the market, both as whole geodes as well as broken up into smaller specimens.

What healers should know Amethyst is common and readily available on the market. Very fine specimens are sold as natural crystals, crystal clusters, or geodes, while lower grade specimens are polished and carved. Usually there is no need to alter the specimens' coloration, but in some cases they may be irradiated to enhance their purple hues. Much like common white quartz, specimens are usually less expensive in rock shops than they are in crystal-healing or alternative medicine shops; amethyst is abundant enough that only the very finest specimens should garner an inflated price. Some large and brittle specimens, such as the famous amethyst cathedrals of Brazil, are repaired by gluing individual crystals onto clusters where they have been damaged, and by reinforcing the external surfaces of the geodes with concrete or another stabilizing substance. Rarely are these repairs disclosed to consumers, however.

NATURAL APPEARANCE

WHAT'S COMMONLY SOLD

 **VS**

A natural cluster of amethyst showing typical variable depth of color

Imperfect crystals are often polished and hand-shaped, such as this one, and may be color-enhanced to increase salability.

Amethyst "cathedrals" prepared for display are ever-popular and thousands come to market each year.

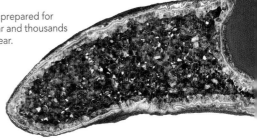

An amethyst "cathedral" still embedded in rock as it was found at a Brazilian mine

Deep-blue coloration in tourmalines, as seen in this example, are among the scarcest and most desirable.

Tourmaline

What they claim Tourmaline, with its various colors, is said to do a great many things, including grounding the user's energy, guarding against negative energy, and aiding during rituals to point to sought-after answers. The protective benefits of tourmaline are supposedly unmatched; it is claimed to be good for defending against everything from ill-wishes and stress to radiation, cell phones, and smog.

What it actually is "Tourmaline" doesn't refer to a single mineral, but rather a family of chemically complex minerals related by their composition and crystal structure. The most common tourmaline mineral, schorl, is typically black and opaque, while elbaite, dravite, uvite, and other members of the series can be more colorful and translucent. All are very hard, brittle, and glassy minerals. Most develop as elongated, striated (grooved) prisms embedded within igneous rocks like granitic pegmatite, or metamorphic rocks like schist. Elbaite is perhaps the most popular, frequently exhibiting good clarity and bright coloration, including crystals showing color zoning. In zoned specimens, the color changes throughout the crystal, sometimes as colored layers along their length, but also as rings of color surrounding each other. "Watermelon tourmaline" is a particularly good example of this, with green tourmaline surrounding a core of pink or red. In general, tourmaline is not rare—schorl is very common in granitic rocks, but is often small and easily overlooked—but very finely formed and colorful crystals are scarce and are typically quite valuable.

What healers should know As with any valuable gemstone, fake tourmaline and color-enhanced specimens are prevalent in the marketplace. Cut tourmaline gemstones should always be approached with skepticism; look closely for flaws, as natural tourmalines virtually always contain some kind of inclusions, flaking, or fractures within them. Great clarity and/or the presence of small bubbles, however, would indicate that the specimen is likely glass. Many whole crystal specimens have also been enhanced through heating or irradiation. Pink tourmaline specimens that have attached smoky quartz, for example, are usually always irradiated (the smoky quartz, made nearly black by the radiation, is a dead giveaway). Lastly, any claims you may have heard of tourmaline's electrical properties actually do contain some truth, but it's far from metaphysical. Due to its internal crystal structure, tourmaline is pyroelectric: it has a natural electric field that is weakened and strengthened by changing temperatures. When heated, tourmaline crystals can draw tiny, lightweight particles toward it, and even attract static electricity, but this attraction fades as the crystal cools again. While it won't protect you from radiation as healers claim, it will keep your hair from getting too staticky when you use a high-end hairdryer containing tourmaline (usually synthetic).

NATURAL APPEARANCE *WHAT'S COMMONLY SOLD*

VS

Perfectly formed and vividly colored tourmalines are rare and often extremely expensive.

While most pink tourmalines are natural, some are color enhanced with radiation. This can turn average-quality, pale specimens into more vivid, salable pieces.

A fine crystal of schorl, the most common tourmaline

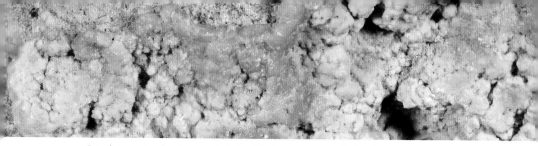
Rough turquoise showing slightly botryoidal (lumpy or grape-like) surfaces

Turquoise

What they claim Turquoise is usually said to be a protective stone—guarding against negative energy, self-harm, and pollution—and one that promotes truth-telling and "wholeness." It is said that it can replenish vitality, stimulate romantic love, and calm mood swings. Some turquoise is even said to be able to treat emotional troubles by "tracing them back along ancestral lines," among other abstract concepts.

What it actually is Turquoise is among the most well-known and popular decorative stones, mined and used for jewelry and charms for thousands of years. Its name comes from the French for "Turkish," not because it was mined there (the ancient source was Persia, or modern-day Iran), but because it was brought to Europe by way of Turkey. The stone gets its trademark color from its copper content, and despite being a fairly tough and hard mineral, it is primarily found in desert regions due to its susceptibility to decomposition in wetter environments. Since it forms primarily as a result of other copper-bearing minerals weathering in acidic conditions, it can be found in almost any kind of rock, often as crusts or pockets in cracks and cavities. These masses are usually opaque and have no shape of their own, instead taking the shape of the space in which they formed; solid, homogeneous chunks are easily cut and polished for use in jewelry. Turquoise crystals are exceedingly rare and found in only a few places around the world. They are usually extremely small, measuring no more than a few millimeters.

What healers should know Turquoise crystals very seldom make it to the marketplace. Most "crystals" found in shops are merely cut and polished chunks of typical turquoise. And as with any popular stone, fakes, modified pieces, and purposely mislabeled specimens abound. Since the most vivid, rich blue colorations are most popular, pale colored turquoise is artificially dyed or otherwise enhanced. Similarly, other minerals of similar hardness and structure to turquoise—but are most certainly not turquoise—are dyed in blue-green hues and sold as turquoise or under trade names like "turquesite." Real turquoise is often crumbly or brittle; many times, specimens are injected with or coated in resin to stabilize them. This is common, accepted practice and does not make them fake, but it should be disclosed to the buyer. Some specimens, however, are composed almost entirely of resin, compacted with real turquoise dust, to simulate the color and texture of turquoise but lack a natural origin; this is called reconstituted turquoise. Lastly, colored turquoise, such as purple or green, is almost always reconstituted turquoise that has been dyed, or even other minerals willfully mislabeled as turquoise. For example, sugilite is a purple mineral often labeled "purple turquoise" (and, ironically, is far scarcer than turquoise), and variscite is a solid green mineral often called "green turquoise."

NATURAL APPEARANCE

VS

WHAT'S COMMONLY SOLD

A vividly colored vein of natural turquoise

While convincing to the untrained eye, this "turquesite" is actually dyed howlite, a mineral that is usually white.

Polished and vividly colored turquoise ready for use in jewelry

Richly colored cubic crystals of fluorite clustered on their host rock

Fluorite

What they claim Fluorite is usually cited as a cleansing stone, introducing positivity in one's life by "sweeping away" negativity, various metaphysical "contaminations" of the spirit, and by warding off psychic manipulation. It is also said to be a stabilizing stone, bringing calm to your emotions, your aura, and your physical body.

What it actually is Fluorite is the most abundant fluorine-bearing mineral, particularly common in sedimentary environments, such as certain limestone formations. It can take a number of forms, quite often as masses or veins within rock, but it also occurs as fine crystals. Fluorite crystals are usually cubic, developing as perfect blocks, but can also be octahedral (an eight-sided shape resembling two pyramids placed base-to-base) or occurring in other shapes. Specimens are very often translucent, sometimes even perfectly transparent, and may be vividly colored in greens, purples, blues, yellows, and occasionally pinks. Some specimens may glow different colors under ultraviolet light, lending to its supposed metaphysical nature, but that is actually a result of the ultraviolet radiation exciting electrons within the crystal lattice of the mineral. While certain colors and crystal shapes are scarcer and more desirable than others, on the whole, fluorite is a very common mineral and easily obtained in the marketplace.

What healers should know As with many minerals that come in a rainbow of colors, fluorite's color can often be altered or enhanced through various amounts of heating or irradiation. Many times these changes are slight or are done well enough that the buyer would never guess they'd been altered, and unless you know what color a specimen from a particular source should look like, you (or even the shop you bought it from) will never know. More common and easier to spot are octahedral "crystals" shaped by hand. Fluorite has perfect cleavage, which means that it will naturally break in certain directions determined by its crystal lattice. In the case of fluorite, its cleavage allows it to break into perfect octahedrons when carefully struck. In this way, octahedral "crystals" can be made from larger crude masses of fluorite. These differ from real octahedral crystals in that their edges are often quite sharp but their faces are rougher and have a slight pearlescence not usually seen on natural crystals.

NATURAL APPEARANCE

WHAT'S COMMONLY SOLD

VS

A natural octahedral fluorite crystal on its host rock. Note the slightly "frosted" appearance.

A fluorite octahedron, carefully cleaved from a larger mass, sold as a natural "crystal." Note the irregular edges and overly lustrous faces.

Rare bluish fluorite crystals

A fine piece of rough lapis from the famous mines at Sar-e-sang in Afghanistan, showing the blue lazulite, white calcite, and brassy pyrite

Lapis Lazuli

What they claim Lapis lazuli is said to be a stone of truth, promoting education and intelligence, while also activating our "psychic centers" and our mind's "third eye." This supposedly stimulates dreams and delivers visions that we may use to interpret our memories and to contact spirits. It is also claimed that lapis lazuli can "aid in our future evolution" by showing us visions of the past.

What it actually is Lapis lazuli's name is derived from ancient Persian and Latin words for "blue stone," and it has been mined and used for decorative purposes for thousands of years. Most specimens on the market originated from the same legendary source in Afghanistan—the same one used by ancient cultures—but export today is limited, keeping lapis lazuli's value relatively high. But lapis lazuli, often simply called lapis, is not a mineral, but a rock, and as such does not form crystals. The various minerals within it can be found individually crystallized, though. These include white calcite, brassy pyrite, and the vivid blues of sodalite and lazulite (which is a varietal name for the blue variant of haüyne, a fairly rare mineral). Worldwide, lapis is rare, having formed in very specific geochemical conditions that lack much silica, or quartz, but it is still common in the marketplace due to its abundance at the few locations at which it is found.

What healers should know Lapis lazuli is a rock and does not form crystals. Any specimens being sold as "lapis crystals" and showing any kind of angular shape have been carved and polished by hand and are not natural forms. In fact, lapis has been carved and polished since antiquity, having been found in ancient Egyptian tombs and sculpture. Sometimes these ancient artifacts come to market, especially small amulets in the shape of scarabs or other Egyptian motifs, and needless to say, good examples are scarce and quite valuable. Though it should always be kept in mind that such artifacts are often the product of antiquities theft and illegal export, and are considered unethical to purchase. Today, lapis is carved into similar Egyptian-styled amulets and figures and usually not sold as ancient artifacts, but the fact remains that some unscrupulous or unwitting sellers may try to pass them off as such.

NATURAL APPEARANCE

WHAT'S COMMONLY SOLD

VS

A vivid example of this ever-popular stone

This is a polished specimen of sodalite, a more abundant, cheaper mineral sometimes passed off as lapis.

Lapis is commonly used in metaphysical products, such as this pendulum made from a color-enhanced specimen.

Though the Afghani lapis is the most famed, it is found in other places. This rare specimen originates from New York State.

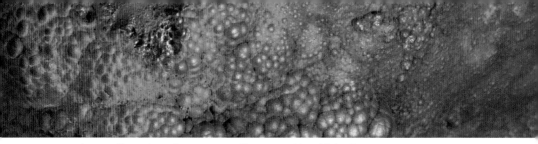

A mass of hematite, an important ore of iron, coated in reddish ochre

Hematite

What they claim Hematite is usually claimed to be one of the strongest grounding stones, keeping body and spirit together while performing metaphysical tasks. The uses for hematite are many; it is often said to "treat" overeating, increase survivability in the wild, and both help to accept and overcome mistakes you've made. It is also said to have a connection to your blood, restoring and regulating blood flow and supply.

What it actually is Hematite is composed of a simple combination of iron and oxygen, and as such is the most common iron-bearing mineral on Earth. It is abundant in almost any geological environment, including soils and muds, and though you may not have found a hematite crystal in nature, you have certainly seen massive or granular hematite as rusty red coatings or stains on rocks. In fact, combined with a similar mineral called goethite, hematite is one of the constituent minerals in common rust. When more finely developed, it takes the form of thin, bladed crystals often in dense clusters, or as botryoidal (lumpy, grape-like) crusts on other rocks or iron ores, both of a dark brown or black color and metallic luster. No matter its color, though, hematite will always produce a rusty red powder when crushed or scratched; ochre, a natural red pigment, consists of compacted grains of hematite. You can usually always distinguish hematite from other similar ores of iron by this reddish powder coloration, its lack of magetism, and its abundance in nature.

What healers should know Hematite is so common in nature and in the marketplace that there is generally no reason to produce fakes or misrepresent it. Polished chunks or broken masses should be inexpensive and unexciting enough that many shops may not even stock it. However, there does exist "magnetic hematite" that is very prevalent in rock shops and even toy stores. This black, metallic substance is virtually always sold as polished nuggets or shaped carvings, but is not hematite or anything even like it. Though the companies that make "magnetic hematite" often keep their exact methods secret, these super-strong magnetic substances are typically composed of a ceramic that contains barium, strontium, and iron. Beware of using this material for healing or any other purpose; the magnetic pull is often so strong that two pieces may slam together abruptly when brought too close, pinching anything caught between them.

NATURAL APPEARANCE

VS

WHAT'S COMMONLY SOLD

This large specimen of hematite shows the rusty red coating and botryoidal (lumpy, grape-like) surfaces this iron ore is known for.

Magnetic "hematite," common in shops, is not hematite at all, despite its similar appearance.

A coarse but finely formed hematite crystal

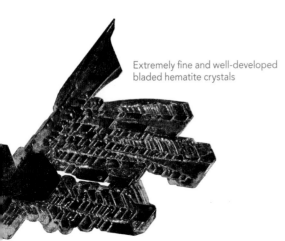

Extremely fine and well-developed bladed hematite crystals

A typical splintery cluster of bladed kyanite crystals, commonly available in shops

Kyanite

What they claim Kyanite is often linked to emotions, said to eliminate fears and anger, promote compassion, and aid in dealing with grief. It is also regarded as able to transfer energy very quickly and never stores negative energy; therefore it never needs to be purified as other stones do. It can also "heal" infections and "treat" various disorders.

What it actually is Kyanite is a common mineral formed primarily in metamorphic geological environments where heat and pressure changed preexisting minerals into new minerals. Containing aluminum and silica, kyanite is an important mineral for determining under what heat and pressure conditions a metamorphic formation developed. If the aluminum-silica mixture had crystallized under a higher temperature, it would have instead become the mineral sillimanite; under lower pressure, it would have instead become the mineral andalusite. All three minerals have the exact same chemical composition, but take on different crystal lattices depending on the external conditions under which they form, so they "record" how their host rock likely formed. Generally, kyanite is black or gray, but the most prized specimens are a vivid blue, and it develops as elongated crystals, often in fan-shaped groupings, that easily splinter along their length. Often whole crystals are difficult to extract from their host rock, and break into small pieces in the process.

What healers should know Kyanite is not as rare as the crystal healing market would have you believe. Vivid blues are less common but are still inexpensive in shops that charge fair prices. Most specimens are merely splinters or fragments of larger crystals or masses; often the broken faces of these specimens show the luster and coloration better than the natural surfaces. These are usually sold as "crystals," however, which is not entirely accurate. Also, considering that andalusite and sillimanite are very nearly the same mineral but formed under different geological conditions, it is curious as to why those two minerals are far less often represented in crystal-healing texts. Both are very common, but most specimens of those two minerals are regarded as less attractive than kyanite, forming blocky crystals with a nondescript coloration, which may be the most likely reason why they don't enjoy the same mysticism as kyanite.

NATURAL APPEARANCE

This specimen shows the splintery nature of blue kyanite.

VS

WHAT'S COMMONLY SOLD

Splinters of kyanite flaked from a larger mass are often polished and sold as natural "crystal wands."

A black "fan" of kyanite crystals

An unusually translucent, gemmy kyanite crystal

A typical rough, broken mass of obsidian with a weathered tan exterior surface

Obsidian

What they claim Obsidian has been called a "mirror stone," due both to its brightly reflective surfaces and its ability to "enhance sight" and clarify one's viewpoints. It is said that these reflective qualities are so effective in revealing truths and making us aware of our faults that it must be used "under careful supervision," and that its power may be overwhelming for some.

What it actually is Obsidian is not technically a rock, nor is it a mineral, and therefore it doesn't form crystals. It is a volcanic glass, and it developed when rhyolite lava was so viscous and dense, and overly rich in silica, that diffusion (or movement) of elements within it could not take place quickly enough to form minerals. When this thick lava met the cool atmosphere, it began to solidify as a solid glassy mass. In some places, particularly in the American West, this occurred with mountain-sized "blobs" of lava, resulting in literal mountains and hills of glass. As such, obsidian is very common and easily obtained; large blocks can usually be purchased relatively inexpensively. Most obsidian is black, opaque in thick masses but translucent in thin sections, occasionally with white spots or brown swirls of color within. It is extremely brittle and breaks into razor-sharp fragments, and when weathered develops a dull, dusty whitish or tan exterior.

What healers should know The marketplace is rife with varieties of obsidian, some real and some invented. As a volcanic material, obsidian frequently contains inclusions of other substances and minerals that mixed in while it cooled; white dots are usually cristobalite (a type of opal), brown streaks or waves of color are usually iron oxides, and "glittering" rainbow iridescence is often caused by tiny crystals of magnetite or other metallic minerals. The metaphysical market likes to find variations of obsidian and give them more fanciful names to increase salability; "mahogany obsidian," for example, is the name for the commonly occurring brown swirls of iron oxides seen in some specimens. Careful cutting and polishing can accentuate otherwise common inclusions, such as the slight rainbow effect often seen in obsidian, therefore increasing its "wondrous" nature (and price tag). Also be wary of green, yellow, blue, red, and white opalescent "obsidians" found in shops; in virtually all cases, these vividly colored stones are just manmade glass.

NATURAL APPEARANCE

VS

WHAT'S COMMONLY SOLD

Called "rainbow obsidian," this desirable variety of obsidian contains microscopic particles of magnetite that produce an iridescent rainbow.

While colored obsidians do exist naturally, they are rare. Colored pebbles sold as "obsidian" are invariably manmade glass.

The famous massive obsidian flow at Newberry National Volcanic Monument, Oregon

Brown-streaked "mahogany obsidian"

A finely crystallized cluster of champagne-colored topazes on their host rock

Topaz

What they claim Topaz is credited with promoting problem-solving, creativity and universal trust in its user, and is said to be able to focus its own energy to the part of your mind, body, or spirit that may need it most. It is reputed to improve your ability to articulate ideas and "clean" your aura and spirit for proper relaxation.

What it actually is A longtime favorite as a gemstone, topaz is a very hard, brittle mineral that comes in a fairly wide range of colors, depending on its environment. It is actually much more common than most people realize, but primarily as small nondescript grains in igneous rocks; well-formed crystals, which developed in cavities, are more scarce. Its crystals are elongated prisms with pointed, wedge-shaped tips that, at first glance, could resemble those of quartz, especially if the topaz is colorless, as is common. Other common colors include yellow, brown, and light green, and less commonly blue or reddish. In ancient times, the name "topaz" likely did not refer to the topaz we know today, but probably to any yellowish gemstone; there is evidence that much ancient "topaz" was actually what we know to be olivine today.

What healers should know In crystal-healing literature, various colors of topaz are often said to perform different metaphysical tasks. Blue topaz in particular is most highly praised among crystal healers, no doubt because of its beauty and subsequent value, especially in the jewelry trade. But real blue topaz is fairly rare, and most is pale in color. Luckily for those marketing the mineral, however, many common brown or yellow specimens can be irradiated to develop a deep, rich blue coloration. This practice is most common in gem-grade stones cut for jewelry, but can be found in the specimen trade as well. For healers, knowing whether or not a blue topaz is natural should be important; since most irradiated specimens began as a pale yellow topaz, would their "healing powers" be that of yellow or blue topaz? Or is the distinction as arbitrary as the healing claims themselves? And, since the deepest-blue specimens are said to work the best for healing, would the holistically inclined crystal healers be onboard with crystals altered by radiation?

NATURAL APPEARANCE

VS

WHAT'S COMMONLY SOLD

River-worn pebbles of colorless topaz are found in various places throughout the world and are fairly common and inexpensive.

After irradiation, colorless topaz crystals and pebbles alike can turn to beautiful shades of blue, more suitable for jewelry and collectors.

A deeply colored (irradiated) blue topaz cut for use in jewelry

A rare and desirable natural blue topaz

A perfectly developed transparent crystal

A glassy fragment of opal showing scarcer reddish-orange "fire"

Opal

What they claim Opal is said to promote optimism and positivity in the user, and to increase vision, both physical and spiritual, with its reflective qualities. It is also credited with increasing spontaneity, faithfulness, and eroticism. But it perhaps sees most use as an amplifier of emotions, as well as a "stealth" stone for remaining "invisible" in dangerous situations.

What it actually is Opal is the name given to several forms of silica (the same material that makes up quartz) that do not have well-organized, regular crystal lattices. As such, opal is not a true mineral and does not form crystals, instead developing as veins or masses that take the shape of the cavity in which it formed. Opal is generally glassy and brittle, often opaque and featureless, as in white specimens, which are known as common opal. But other varieties, called gem opal or precious opal, are more translucent and show a colorful opalescence; this occurs because of the way light enters and bounces within the stone. In the finest specimens, this results in an internal "fiery" play of light and color. These beautiful varieties of opal are undoubtedly the most valuable and sought after—especially from famous sources, such as Boulder Opal from Australia and Fire Opal from Mexico—but are far scarcer than common opal. However, all opals suffer from a similar habit of crazing, or developing fine cracks on their external surfaces, when exposed to air and heat too long. Many collectors store their stones in mineral oil in an attempt to prevent this.

What healers should know Opal, like obsidian and amber, is yet another stone popular with crystal healers that does not form crystals and therefore does not follow much of the crystal-based healing jargon, but is repeatedly recommended anyway. Many "crystals" on the market are merely cut and polished masses. Precious opal can be very expensive if the internal play of color is particularly colorful and bright, and as a result, many lower-grade specimens are enhanced in an attempt to elevate their value. In jewelry, many low-grade opals are fashioned into what are known as triplets. An opal is cut very thinly and glued to a backing material, often a black stone, which gives the opal more contrast. A quartz cap is glued to the top of the opal, which both protects it and takes a better polish, making the opal appear more lustrous. This has long been a common practice, but many folks buying a piece of opal jewelry are unaware that their low-grade opal has been altered. Similarly, some opal on the market is outright fake; synthetic opal is fairly easy to make, even with colorful "fire" within, and opalescent glass has been used as an opal substitute for decades.

NATURAL APPEARANCE

VS

WHAT'S COMMONLY SOLD

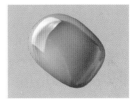

Most common opal appears as glassy white masses. The colorful "fire," if present at all, is often isolated to small areas in the stone.

Often presented as an inexpensive form of opal, "opalite" is actually manmade glass.

Common white opal embedded in its host rock

Spectacular "fiery" veins of colorful opal in their host rock

Amazonite is commonly sold cut and polished.

Amazonite

What they claim Amazonite is said to promote courage and confidence in its users, allowing them to manifest their goals and dreams. It is also said to be particularly powerful in absorbing negative environmental energies, such as radiation, microwaves, and "electrical smog"; it is said that it should be placed between your cell phone and your body at all times. It is also purportedly a stone of truth, and any visions received while wearing it should be trusted.

What it actually is Amazonite is the varietal name given to blue-green examples of microcline, one of the most common feldspar minerals. Most famous from Colorado and Brazil, its color has been found to be the result of lead impurities, in most cases. Like other feldspars, it develops as elongated, blocky crystals with angular tips, and particularly fine specimens may show preferential coloration: coloration where some crystal faces are more richly colored than others. In Colorado, famous localities for amazonite yield specimens associated with smoky quartz, with crystals of both minerals intimately intergrown to make for beautiful and very valuable specimens. Lower-grade specimens with poor crystallization are often broken up and polished for use in jewelry or sold as inexpensive collector stones—many of these originate from Brazil and are more green in color, often with short white streaks throughout. In all cases, amazonite is fairly rare worldwide, but the localities that produce it do so in large quantities.

What healers should know Despite being named for the Amazon River, there is little, if any, historical evidence that this colored feldspar was ever found in its vicinity. Though many tons of amazonite have originated from elsewhere in Brazil, the greenish stone historically found along the Amazon River was likely of a different mineral entirely, though its identity is now lost. Whatever the case, today there is no amazonite found in the Amazon River, and any such claims in the marketplace are false. And as with many colorful stones, low-grade amazonite or other feldspars may be dyed or color-enhanced to improve salability, though tumble-polished specimens are common enough on the market that this usually isn't necessary. The beautiful crystal clusters from Colorado, however, garner such attention and high prices that some concocted specimens exist. Multiple crystals may be carefully glued together or have attractive smoky quartz added in order to increase their salability. Thankfully this practice is not very common, but if you plan to invest in a valuable specimen, great care should be taken when inspecting the piece.

NATURAL APPEARANCE

WHAT'S COMMONLY SOLD

VS

A typical example of rough, natural amazonite of good, rich color

Occasionally, other feldspars or other minerals entirely are dyed green-blue to resemble amazonite and are sold as such. This is dyed quartz, advertised as amazonite.

A vividly colored and well-formed cluster of amazonite crystals from Colorado showing preferential coloration on crystal faces

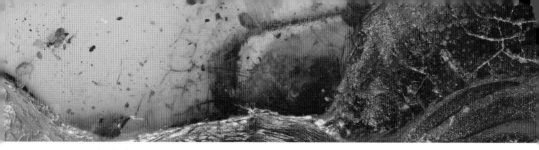

A chunk of rough amber

Amber

What they claim Amber is said to promote a sense of well-being by introducing warmth into one's life. As a result, it promotes comfort, especially when one is ill or injured, and can supposedly clear depression or sorrow. It is also said to be a motivating stone, able to help accomplish tasks and make clear decisions.

What it actually is Amber is fossilized tree resin, hardened through the process of polymerization (the organic material in the resin begins to combine into solids). This process is facilitated by the pressure of being buried underground, which can keep the resin concentrated and protected from decomposition. The result is a hard, translucent to transparent substance, often of golden yellow or orange coloration, which has been used as a gemstone for millennia. In the Baltic region of Europe, nuggets and chunks of amber can occasionally be found on beaches or along rivers, while in most other locales must be mined. But what makes amber most compelling is the fact that it can contain traces of ancient life trapped within it, such as small insects, spiders, leaves, flowers, feathers, and even entire lizards that were unlucky enough to get stuck in the sticky resin as it seeped from a tree. Amber ranges wildly in age; some specimens are only a million years old; others formed some 300 million years ago. Some sources, such as Colombia, produce a similar substance, called copal, which is young resin that has not hardened enough to be considered true amber, yet is often marketed as amber.

What healers should know Amber is a perennial favorite, treasured since antiquity for its color, clarity, and fossil inclusions. Baltic amber, from Northern Europe, considered to be the finest amber, is becoming more scarce, yet demand, especially in Asia, has increased. This has resulted in dangerous and illegal mining, particularly in Russia and Ukraine, where the mafia have become experts in rapidly obtaining amber. When a potential amber-rich site has been identified, often on protected or sensitive forestland, miners quickly clear-cut and remove the trees, then use high-pressure hoses to blast through the soil in search of buried amber. Often heavily armed and warring for the best sites, these illegal miners quickly take what they can and then abandon the area. In the Dominican Republic, amber mining is a rural affair, done in jungles and hillsides with hand tools. Often the mining is done in small pits prone to filling with water and collapsing during sudden rainstorms, suffocating miners. Though not all amber is obtained in such dangerous ways, enough of it is that one must consider the social ramifications when buying it. Lastly, there are countless fakes on the market; amber-colored resins, often with insects inserted by hand, are cheap, easy to produce, and convincing.

NATURAL APPEARANCE

WHAT'S COMMONLY SOLD

VS

Rough water-worn amber often doesn't show much clarity until broken or cut open, but internally may have variable textures and inclusions.

Artificial amber, made from resin, is often sold pre-cut and polished for use in jewelry. Much fake amber is replete with little flakes and other internal variations that make it "sparkle" in the light

A tiny ancient fly, about ⅛-inch long, trapped in a polished specimen of amber.

A diamond crystal still embedded in its kimberlite host rock

Diamond

What they claim Much like their incredibly hard nature, diamonds are said to promote tenacity and toughness, both mentally and physically. It is said that diamonds can eliminate pain and can be used as an "amplifier stone" that increases the effectiveness of other stones.

What it actually is Like gold, copper, and sulfur, diamond is a native element and is one of the crystallized forms of pure carbon (the other is graphite). It is the hardest naturally occurring mineral, and can be subjected to great heat and pressure without it changing—though diamonds are very brittle and can be chipped fairly easily. Usually colorless unless impure, diamonds crystallize in cubic or octahedral shapes and are prized for their "flashy" luster, which can be accentuated by cutting diamonds at specific angles. While they can be found loose in rivers, most diamonds are mined directly from host rock—most frequently a peculiar type of rock called kimberlite. Kimberlite forms as tall cylindrical columns that rise through other types of rocks, formed during explosive eruptions from magma chambers deep within the earth. When healers insist that "crystals form deep in the earth," diamond is one of a handful of minerals for which that description is actually accurate. The diamonds within kimberlites already existed when the kimberlite eruption blasted them upward.

What healers should know Diamond's rise to popularity and high valuation are as much the results of marketing as they are of diamond's natural beauty and relative rarity. Once valued less than ruby, sapphire, and other colored stones, diamonds enjoyed a surge in popularity—particularly in the early- to mid-twentieth century in North America. This ascent coincided with the rise of Hollywood, product placement in films, and publicity surrounding celebrities. Clever marketing positioned diamonds as engagement stones and persistent reinforcement of the concept solidified diamond's value and popularity in today's jewelry market. But with great popularity and value come great social and human rights concerns. Conflict diamonds, also called "blood diamonds," are diamonds mined by revolutionary groups, criminal organizations, or other violent, conflict-driven groups that protect their mines by murdering opposition, use slave labor, and/or employ other unethical means to acquire diamonds. Much attention has been brought to this issue in recent years, and agreements and regulations have been enacted in an attempt to curtail the spread and sale of conflict diamonds. Nonetheless, conflict diamonds are still very much a problem and are available on the market, making the purchase of a diamond without verification of its source irresponsible.

NATURAL APPEARANCE

VS

WHAT'S COMMONLY SOLD

This diamond, still partially attached to its kimberlite host, shows the typical crystal structure of natural diamonds.

Most shops know better, but a type of crystal called "Herkimer diamond" may be passed as a diamond. Named for its locality in New York, these are just very clear quartz.

A well-developed diamond crystal of typical gray-brown coloration

In well-developed labradorite specimens, you can clearly observe the finely layered structure that produces the shimmering effect.

Labradorite

What they claim Labradorite, with its colorful schiller effect, is considered among the most mystical of stones, bringing light and protection to its user. It is said to deflect negative energies and protect our inner energy reserve. It is also claimed to be an important and powerful stone for telepathy, reading the future, spiritual travel, and communication with spiritual beings.

What it actually is Like moonstone, sunstone, and amazonite, labradorite is a type of feldspar, the most abundant group of minerals on Earth. Technically, it is a variety of the feldspar anorthite, but it also contains feldspars of differing composition, which formed in tight, often microscopic layers. Much like in other iridescent or pearlescent materials, when light enters and bounces between these varying layers, it produces a blue-green to yellowish (and rarely pink) schiller, an optical phenomenon known as labradorescence. The intensity and color of the schiller changes as the viewing angle of the stone is changed, creating an attractive, variable appearance. In very well-formed specimens, the lamellar, or layered, structure can be easily seen; most specimens, however, are small grains embedded in igneous rocks like gabbro. Labradorite is named for Labrador, Canada, a historic source of fine specimens, though most specimens on the market today originate from Brazil, Finland, and especially Madagascar. Not all specimens show the trademark labradorescence, however; much labradorite is dark and opaque, while particularly clear, gemmy examples may actually be too translucent to produce the schiller effect.

What healers should know Madagascar is a major source for gem-grade labradorite, but much like other minerals mined on the island, it is largely obtained in sensitive environments by rural workers. Safety regulations may be lax or poorly enforced, and miners may be underpaid for the hard, dusty, hot work. Child labor is another real concern. In addition, due to a perceived greater value in specimens originating from its namesake Labrador, Canada, sometimes specimens from Madagascar or other countries are labeled as Canadian, due in equal parts to ignorance and deceitful marketing. Much of the time, this is not necessarily the fault of the retailer, as source information is often lost or disregarded before the material makes it to consumers.

NATURAL APPEARANCE

WHAT'S COMMONLY SOLD

VS

Even without cutting or polishing, the finest labradorite specimens will show a strong schiller even when rough.

The variety of labradorite that exhibits its trademark play of color does not form fine crystal points. Any specimen like this has been shaped and polished by hand.

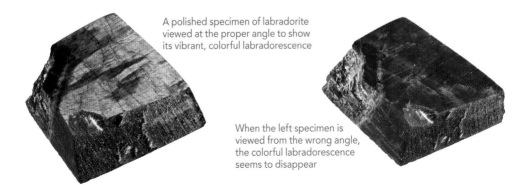

A polished specimen of labradorite viewed at the proper angle to show its vibrant, colorful labradorescence

When the left specimen is viewed from the wrong angle, the colorful labradorescence seems to disappear

A rough piece of Himalayan salt

Himalayan Salt

What they claim Pink Himalayan salt is said to provide all manner of health benefits, such as greater nutrition when consumed and detoxifying effects when in its presence. In decorative form, often hollowed and used as a lamp, it is said that the salt releases negative ions into the air which can rid a room of airborne dust and other pollutants.

What it actually is Himalayan salt is, as its name implies, salt from the foothills of the Himalayas, known for its distinctive orange-pink coloration. Salt is mineralogically known as halite; it is a common combination of sodium and chlorine, usually deposited as an evaporite mineral left behind when salty seawater dries up. It is often associated with gypsum, sylvite, and anhydrite. The colorful halite usually marketed as Himalayan salt originates from the Punjab area of Pakistan, in a body of rock appropriately named the Salt Range Formation, not all that far from the western end of the Himalayan Mountains. This massive bed of salt was deposited approximately 550 million years ago, and has been a global source of salt for centuries. Much of the production is done at the Khewra Salt Mine, said to have been discovered by the army of Alexander the Great, which yields nearly 400,000 tons of halite per year today for decorative, culinary, and metaphysical uses. Halite is a common mineral worldwide; when finely crystallized, it forms transparent cubic crystals, but salt beds are usually dense, opaque, and mixed in with other evaporite minerals. In all forms, salt is quite soft and easily dissolves in water.

What healers should know Himalayan salt has been the subject of masterful marketing tactics in recent years; the chief selling points are that it's healthier than common table salt or that it can purify the air in a room. Its trademark orange-pink coloration is caused by trace impurities of other minerals, which many people insist make it healthier to consume. In reality, multiple studies have shown it to be no healthier than common table salt. In fact, it's less healthy as it lacks the biologically essential element iodine, which is commonly added to table salt. In addition, at some sources, the coloration in halite is caused by insoluble or sometimes toxic mineral impurities, usually in concentrations so low that they are safe enough to consume, but still worthy of note. Also common on the market are Himalayan salt lamps, in which a chunk of salt has been hollowed out and had a light bulb placed inside; the light then glows through the translucent salt. These are said to do everything from purify the air in a room to calm those nearby, but there is no scientific evidence that the heat or light of the lamp reacts with the halite in any way. In addition, because halite is so susceptible to dissolving—the surfaces of a specimen may begin to melt just with the humidity in the air—lamps and other decorative pieces are often coated in resin or plastic to protect them, sealing them off from the atmosphere and most certainly blocking any supposed metaphysical benefits.

NATURAL APPEARANCE

WHAT'S COMMONLY SOLD

VS

Much Himalayan salt has natural variations in color, caused by impurities of other minerals.

You will most likely encounter so-called Himalayan salt more frequently for culinary applications, often cited as being more healthy than typical salt.

A more typical, less vibrant moldavite specimen

Moldavite

What they claim Moldavite is said to allow us to communicate with our spiritual selves, and gazing into it apparently can alter your consciousness and facilitate "ascendance." It is generally regarded as a very powerful stone and can be used to "go back and forth through time" or even to "inhabit other lives" for a short time. Moldavite is also claimed to help "star children," or people who do not feel at home on Earth, to alleviate their feelings of being out of place.

What it actually is Moldavite is a particular variety of tektite, or impact glass, found primarily in the Czech Republic and, to a far lesser extent, Germany. It formed when a meteorite struck the earth around 15 million years ago in what is today Bavaria, Germany; this ejected large amounts of pulverized and molten rock into the atmosphere. As the droplets of molten rock quickly cooled in the air, they had no time to develop any crystals and instead hardened as glass before they even hit the ground. This is the process by which all tektites form, but the rock that was melted to become moldavite was of such a mineral composition that the resultant aluminum- and silica-rich glass developed a distinctive forest-green coloration. Other qualities typical to moldavite include its translucency, often with few impurities, and its "wrinkled," "veiny," and pitted surface texture, developed as the air channeled along its surface as it fell. As a result of its means of formation, most samples are rounded, oblong, and usually fairly thin and were shaped as they spun through the air.

What healers should know Most sellers of moldavite, particularly retailers of metaphysical goods, will emphasize the finite quantity of available specimens. As it formed from a single meteor impact, it was theoretically produced only one time, and therefore specimens will run out some day. While this may be true, thousands of specimens are still discovered each year, and new sources could still be discovered at some point. But that doesn't stop sellers from asking astronomical prices for nice specimens today. Many of the most valuable specimens are those that developed atypical shapes and structures, different from the usual rounded tektite "blobs." While many of these are perfectly natural, others have been shaped by hand to emphasize or create more desirable shapes. As moldavite is a glass with no crystal structure, it can be difficult to distinguish an altered specimen from a natural one. Similarly, manmade green glass can be molded, etched, and carved in such a way as to mimic natural moldavite, and an untrained eye may not be able to tell the difference. Lastly, there are few legal mines of moldavite in Czechia; lots of moldavite is obtained illegally by digging on privately owned land, often under cover of night.

NATURAL APPEARANCE

WHAT'S COMMONLY SOLD

VS

Typical moldavite specimens are dark yellow-green with rounded or "blob-like" exteriors with hazy or "frosted" surfaces.

Fake moldavite is common in the marketplace, especially as cut gemstones. This piece, advertised as moldavite, is far too green and too flawless; it is manmade glass.

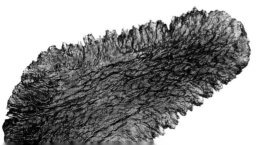

An extremely fine specimen of moldavite, with rich coloration and good clarity. The sculptural aspect of its edges make it a particularly valuable specimen.

ALTERED, FAKE, AND INVENTED "CRYSTALS"

With thousands of known minerals and hundreds of distinct rock types, you'd think there would be plenty of species to use in metaphysics without needing to invent more. But when most of the commonly available rocks and minerals are abundant and inexpensive, they leave much to be desired in the eyes of marketers and metaphysical retailers. For no other reason than increasing profit margins, many common rocks and minerals are treated, colored, and given creative names to deceive consumers into thinking that they are something new, exciting, and have greater "healing" properties. In some cases, the creators or "discoverers" of these "new" materials may be so entrenched in their metaphysical beliefs that they actually think that they have found something special, ignoring scientific research and testing that shows their find to be something well-known and often as common as quartz. Many times the names concocted for such materials are trademarked by the individual or company that devised it, in an attempt to both protect and perpetuate their claims as legitimate. As a trademark is designed to protect intellectual property, the presence of a trademark symbol after a supposed mineral name should be your first clue that whatever claims that follow it are fictional and have no basis in accepted science. The following are just a sampling of the dozens of examples found in the marketplace today.

"Aqua aura" quartz

"Aqua aura" and "titanium quartz" These two varieties of colorful quartz are altered to make common quartz crystals more salable. "Aqua aura" is quartz that has been treated in a lab with minute amounts of gold that create an iridescent shimmer on its surfaces, while "titanium quartz" has been coated with titanium or other metals to give it opaque, metallic surfaces. Both are just common, transparent quartz beneath their manmade exteriors.

"Rainbow quartz" Whether in crystalline or tumble-polished forms, "rainbow quartz" is dyed quartz. Often shocked by rapidly cooling crystals after heating them to high temperature, these crystals are full of small cracks that better hold the colored dyes, usually in shades of blue, purple, pink, and green.

Dyed "rainbow quartz" crystals

"Merlinite" Often cited as a particularly powerful healing stone, "merlinite" is an invented name given to dendritic chalcedony. Chalcedony, composed of microscopic quartz crystals, can sometimes contain small dendrites, or tree-like growths of other minerals within it; this is a fairly common occurrence, found in many places around the world. "Merlinite" is simply a trade name for an otherwise abundant stone.

"Merlinite"

"Super seven," also called "Melody's stone" This material supposedly contains seven different minerals that combine to make it a very powerful healing stone, but the entire concept is flawed from the outset. Three of the minerals said to be in Super Seven are amethyst, smoky quartz, and quartz, all of which are the same mineral. Rutile and goethite are usually needle-shaped minerals that are said to be present, and are common

enough in quartz that many times they may actually be, but two much scarcer minerals, cacoxenite and lepidocrocite, are rarely if ever present in a typical Super Seven specimen (most "cacoxenite" in these pieces is simply misidentified goethite). Some healing books also cite hematite as one of the seven minerals, as well. As pretty as some polished specimens of this material may be, the fact is that it is just impure, colored quartz, and would have likely gone unnoticed in the marketplace if marketers had not spun a story around it.

Trademarked stones When reading crystal-healing literature or shopping alternative medicine websites, you will likely come across materials with trademarked names. Azeztulite™ and Rosophia™ are two prominent examples. With these stones, the discoverers "felt" something different about them and began marketing them as particularly powerful metaphysical stones. All tests and observations have shown North Carolina's Azeztulite™ to be common white quartz and the Rocky Mountains' Rosophia™ to be a granitic rock rich with orange-red feldspars and white quartz; both are extremely abundant

Azeztulite™ is quartz from North Carolina

Rosophia™, a reddish granite from the Rocky Mountains

materials of no particular interest. But that hasn't stopped the inventors of these names from protecting their claims by trademarking the stones and crafting successful businesses around their supposed metaphysical properties.

❝Perhaps most alarming of all, some crystal-healing books promote contact with radioactive minerals in a dangerous and misguided attempt at providing 'cures.'❞

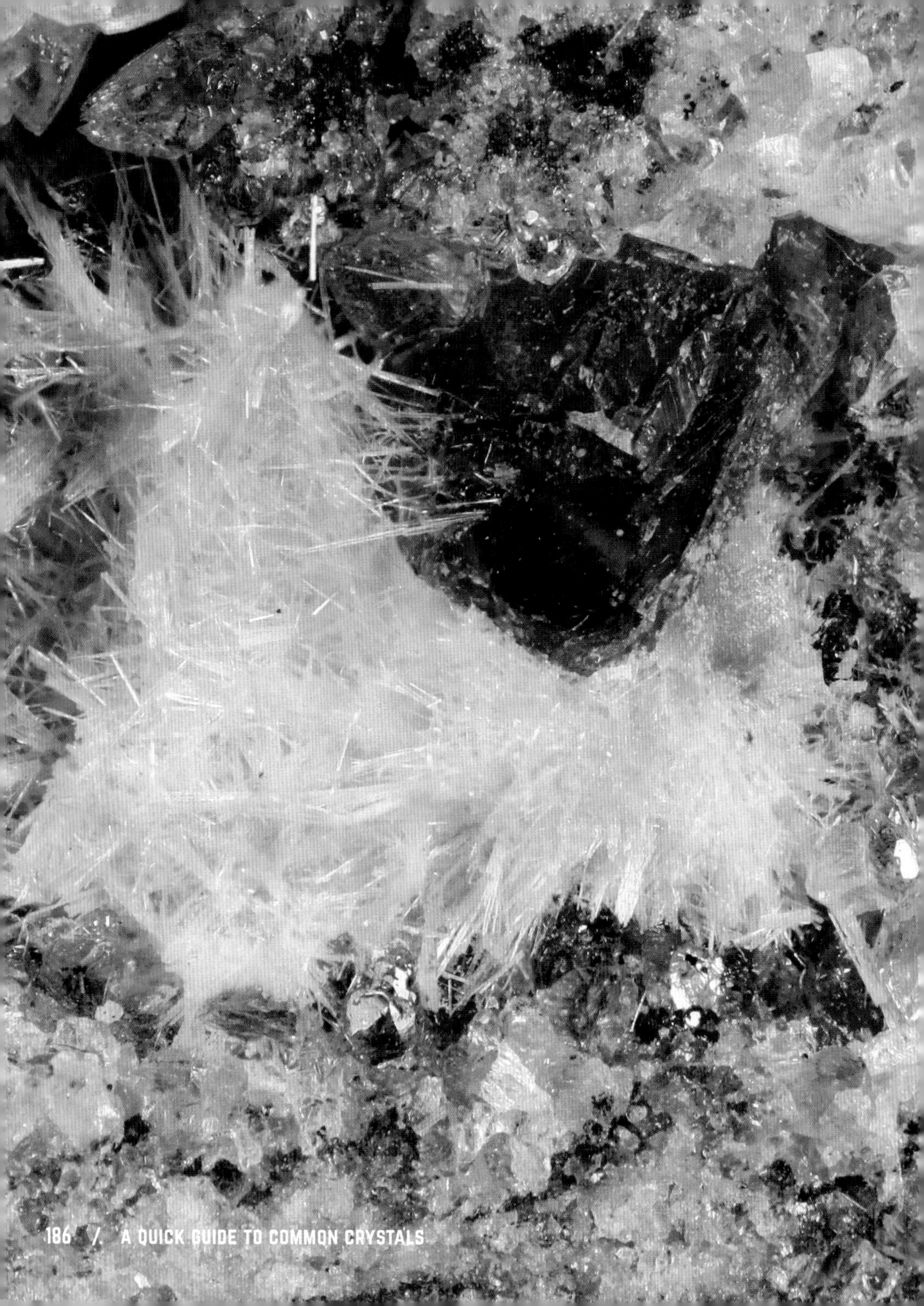

UNFORTUNATE OMISSIONS

It should be noted that throughout crystal-healing literature, little mention is made of any minerals that contain harmful elements. Radioactive minerals are absent from most crystal-healing texts, and minerals bearing elements like lead, arsenic, and mercury are not often included, either. This is surely because the authors do not want to encourage contact with potentially harmful minerals. But if they are not included, if only to warn readers, then what is to stop an unaware amateur crystal healer from handling or consuming one of these minerals on their own? (After all, they're often highly collectible and easy to find online.) A few books do include minerals that contain harmful elements but make little effort to warn their readers or advise against their use, which is as problematic and irresponsible as not including them at all. For example, in one book, cinnabar, which contains mercury, is included, but there is no information on how to handle it safely, what precautions to take, or how to safely store it.

SCIENTIFIC SIDEBAR

While uncommon, a few crystal-healing books include radioactive minerals as "healing" crystals. Consider uranophane, a radioactive mineral that develops as tiny yellow needle-like crystals. To some authors, it is reputed to be a "homeopathic" cure for "radiation damage." This is a deeply problematic concept and promotes contact with a radioactive substance, particularly one with small, sharp crystals that, if handled without gloves, can become embedded in your skin. This an incredibly dangerous and misguided notion. Under no circumstances should you use uranophane or any other radioactive minerals for "healing" or any other alternative healing purpose. Even knowledgeable collectors and mineralogists take radioactive minerals seriously, storing them away from food, water, and living areas, keeping them ventilated to avoid build-up of toxic radon gas, and limiting exposure in general, no matter how low the radiation risk.

OPPOSITE: A radioactive cluster of yellow needle-like uranophane crystals

THE PROBLEMS OF CRYSTAL HEALING: A SUMMARY

Crystal healing, the messaging and marketing that surrounds it, and New Age metaphysical ideals on the whole have repeatedly shown themselves to be problematic aspects of today's alternative lifestyle culture. Believers often preach open-mindedness while disregarding and actively rejecting scientific fact; promote oneness with and respect for the earth while simultaneously driving a booming crystal-mining industry; and claim to have a deeper understanding or inner knowledge of crystals while choosing to ignore the (amazing and truly awesome) natural processes that produce crystals. But while hypocrisy and misinformation are the bywords of crystal healing, they aren't the only problematic aspects. As we've shown throughout this book, there are a number of ways in which the crystal-healing industry has proven harmful. They fall into three main categories.

1. CRYSTAL HEALING IS A MARKETING-DRIVEN, PROFIT-FUELED INDUSTRY

- The New Age metaphysics industry exhibits questionable ethics by marketing crystal healing as a miracle cure and a panacea for all manner of physical and mental ailments, including as-yet incurable diseases. This often preys on the vulnerable when they need help the most.

- Crystal healing takes advantage of its target markets by making spurious claims to curious individuals seeking alternatives to modern medical treatments, and by preying on individuals who are interested in minerals and crystals but often don't know much about them.

- The New Age metaphysics industry willfully promotes misinformation, false claims, and anti-science ideals that are potentially harmful and costly to those who are convinced to believe them; in place of a scientific approach, they promote archaic mysticism and magic.

- Marketers promote the largest, clearest crystal specimens as the "best" for healing in an attempt to justify the gross pricing markup found at New Age crystal-healing shops; they also concoct (and trade-mark) creative names and stories around common, often low-grade minerals in order to market them and charge exorbitant prices.

- Crystal healing is rife with hypocrisy, regarding crystals as "gifts from nature" and seemingly promoting a holistic, environmentally in-tune, anti-big-business movement while simultaneously trading specimens as commodities and driving a multi-billion dollar industry predicated on mining often in developing countries where mines are unsafe, illegal, or do great damage to the environment.

2. SCIENCE DOES NOT SUPPORT THE CLAIMS OF CRYSTAL HEALING AND METAPHYSICS

- Crystal healing cannot stand up to the scientific method; it is not supported by evidence-based observations, replicable peer-reviewed testing, or falsification of its claims.

- Inconsistencies among crystal-healing authors and practitioners show that each has a different experience with crystal healing, high-lighting its highly subjective nature; science strives to remain as objective as possible, building upon previous knowledge and research rather than coming up with one's own interpretation.

- Crystal healing often disregards well-understood scientific findings in favor of its own baseless interpretations (e.g., that crystals are "mys-terious beings" rather than molecular patterns), or may take complex scientific phenomena poorly understood by the general public and appropriate the language/jargon as support for their claims (e.g., citing quantum physics as an answer for how crystal healing works), while being unable to sufficiently explain how the theories they are appropriating link to supposed crystal healing.

- Claims made by crystal healers are often expected to be accepted as fact based on faith in crystal healing alone. Such claims are supported only by unverifiable anecdotal evidence and misattribution of the perceived benefits of alternative healing.

- New Age metaphysicists make no attempts to prove themselves wrong or falsify their own claims, and everything is presented as "correct" unless they are challenged, in which case they are able to provide many dubious reasons why the challenger is incorrect.

- Crystal-healing books often position themselves as "encyclopedias," "bibles," or "definitive guides" to crystals while offering little to no actual scientific or authoritative information about crystals and minerals.

"Minerals, flowers, essential oils—these are some of the typical trappings of metaphysical practices."

3. UNINTENDED CONSEQUENCES OF BELIEVING IN CRYSTAL HEALING CAN CAUSE HARM

- By perpetuating untestable, unverifiable folk-healing concepts in popular media, crystal healing becomes legitimized and more people may come to believe its unscientific claims.

- Using crystals to "treat" illness or as a substitute for proper medical treatments, such as vaccines, could compromise personal or public health and safety as curable illness and disease are allowed to persist longer than necessary.

- Misguided attempts to use crystals and minerals in inadvisable ways (e.g., drinking mineral infusions) could lead to poisoning or death.

- Crystal-healing books often don't mention toxic minerals in their books. This is a dangerous omission that could lead to confusion in their readers and followers. For example, an author may choose to omit a lead-based mineral due to its potentially harmful nature. But a reader may find such mineral in a shop and, upon not finding it in the book they consult, buy it and attempt to use it for healing anyway, unaware of its dangers. (As it happens, many minerals containing toxic elements, such as lead and arsenic, are often beautiful and popular collectibles).

- Many crystal-healing techniques are potentially unsafe if performed with the wrong minerals; promoting their use with flowery and scientifically uninformative language is irresponsible.

- The crystal-healing industry perpetuates the mining of mineral specimens, some of which is done illegally, in sensitive environments, by poorly trained and underpaid rural workers, often in dangerous conditions, and often with destructive or polluting mining techniques.

- Individuals who may be genuinely interested in the natural aspects of minerals may first encounter crystals through the unscientific lens of crystal healing and are thereby misled and miseducated from the start.

- Crystal healing unethically takes advantage of those who have limited access or funds to seek traditional hospital care and modern medicine, and offers false hope to those who have reached the limits of what modern medicine is able to treat.

These are just a few of the problems with crystal healing, but there are a few themes that connect them: faith and insistence in a New Age metaphysical worldview that promote a certain variety of ignorance. These claims are not rooted in reality, but crystal healing continues to exist because the marketing around it is so successful. Social media spreads these unchecked problematic ideas faster than ever, and when unburdened by the scientific method, crystal-healing authors can be prolific and produce title after title.

Science, on the other hand, is an inherently slow and often confusing process, full of complicated research and consideration for all aspects of the topic, both for and against it. But it is an important process that roots out fact from fiction, anecdote from evidence. While it can be dissatisfying to learn that science hasn't solved some of the biggest health problems we face today, improvements surround us. Chemotherapy and cancer survival rates, for example, are vastly improved today over just a couple decades ago. Similarly, a great deal of progress has been made in terms of treating and preventing HIV infections, a notoriously difficult disease, though the affordability of drugs does admittedly remain a challenge.

Difficulties of science aside, the simple quick-fix "answers" provided by New Age metaphysics shouldn't be your next option. By employing evidence-based thinking and choosing to gather as much information as possible from both sides of an argument before believing in either, you can become a more active and engaged thinker. This will also make you less susceptible to marketing and more receptive to complex ideas. For some, the narrative created by New Age metaphysics creates a fulfilling ideology. But for the rest of us, we can all learn to be more critical thinkers simply by taking a step back from a crystal healing claim and asking ourselves, "What do they get out of this?"

Though undoubtedly beautiful, exhibiting some of the most vibrant and exciting colors in the crystal world, cinnabar (below) and realgar (to right) contain toxic elements. Though these two minerals are often irresponsibly recommended in crystal-healing books, the mercury in cinnabar and the arsenic in realgar make it inadvisable to come into contact with either mineral, especially for crystal-healing practices.

SCIENTIFIC SPOTLIGHT: EVIDENCED-BASED REPLACEMENTS TO CRYSTAL HEALING

Stress and relaxation techniques Practices such as guided imagery, breathing exercises, and muscle relaxation can help individuals cope with anxiety brought on by chronic health problems, like heart disease and inflammatory bowel disease (IBS). Stress and relaxation techniques practiced before bedtime can lead to improved sleep quality, too.

Meditative movement Practices include both meditative and physical elements, such as yoga and tai chi. Yoga, rooted in Indian philosophy and spirituality, has become popular in Western culture as a way of promoting physical and mental health. Although more research is needed, studies suggest that yoga relieves pain, reduces stress, and improves sleep and balance.

Social support Positive and meaningful connections with others are considered fundamental to mental and physical health. Socially supportive relationships are consistently correlated with better biological functioning and lower risk of mental and physical illness.

Hobbies Engaging in a satisfying activity that you find enjoyable, such as rock and mineral collecting, is also beneficial to mental and physical health. Though the scientific research is still relatively sparse, one study has so far shown that older adults who took part in meaningful productive activities, like gardening or volunteering, lived longer than people who did not. People who volunteer in their communities also report feeling happier than their peers who are less active. Participating in the creative arts is also showing to be a promising way to enhance cognitive function, including improving memory, comprehension, and problem solving. Additional clinical and experimental studies are needed to determine what exactly is causing these improvements.

It is important to emphasize that these practices are still being evaluated by the scientific community. Though they may be helpful for addressing certain aspects of mental and physical health issues, they should not replace formal medical treatment for diagnosed medical conditions, like cancer, heart disease, or mental health disorders. If you have concerns or would like to learn more about what you can do to take better care of your mental and physical health, the best strategy is to visit your primary health physician or seek the help of a therapist who can give you guidance on what steps you should take.

When starting a mineral collection, it can be helpful to have a collecting theme. Many people enjoy collecting minerals used as gemstones, for example, such as this vivid, well-crystallized spinel.

Another fun way to theme a mineral collection could be to focus on minerals not found in crystal-healing books, such as this metallic, magnetic pyrrhotite crystal. This lesser known mineral is too obscure for most crystal healing authors to include, so do they still believe that it has healing properties?

SHOULD YOU PRACTICE CRYSTAL HEALING?

It's conclusive that crystals do not and cannot impart any metaphysical energy to their holder, but that doesn't mean that you can't still find enjoyment with crystals. While the crystal-healing industry is rife with misinformation and problematic practices, rock and mineral collecting in general can be done safely and ethically, and the real benefits of the hobby may exceed those provided by crystal healing. In this chapter, we'll summarize the issues with crystal healing and give you an idea of better ways to manage your health.

"Crystal healing isn't inherently harmful, of course. The harm comes from the problematic ideas and anti-science culture that surrounds crystal healing."

CRYSTAL HEALING, IN CONTEXT

The claims made by crystal healers are unsupported by science and go against accepted facts; crystals don't contain earthly magic and certainly can't heal you. In fact, some may actively harm you. But for some people—even those with a level-headed understanding of how crystal healing does and doesn't work—there may still be a draw to practice some of its techniques. And if the concept still appeals to you, that's OK! Feel free to give some of the relaxation, meditation, and mindfulness-based crystal-healing techniques a try, but do so safely with non-toxic minerals and a clear understanding of where any benefits you may feel are really coming from. If you feel that practices like these could reduce stress and provide a sense of enjoyment, then it may be worth a shot. Just remember that while meditation, for example, has been shown to have measurable benefits, they are derived from the acts of deep breathing and calming your thoughts, not the crystals in the room.

In most cases, crystal healing or crystal therapy isn't inherently harmful, of course. The harm comes from the problematic ideas and culture that surround crystal healing—the anti-science, pro-myth gurus who promote their own books, websites and a gemstone industry under the guise of offering a more holistic and "authentic" lifestyle. The proliferation of false information and an insincere, often hypocritical altruism shown by metaphysical writers serves only to reinforce the fiction they sell. But by learning how to see through the marketing and mysticism, using evidence-based reasoning and critical thinking to navigate the crystal healing jargon, you can protect yourself from the pseudoscience and enjoy minerals for what they are, not what they are said to be. As long as you've established realistic expectations for what the practice is actually doing, you may come to enjoy meditation with stones without the need for magical claims; the "healing" you may feel is simply a product of doing a thing you find enjoyable or relaxing.

There are also those who engage with crystal healing because they have an interest in rocks and minerals, but the two activities don't need to go together. If you enjoy rocks, minerals, and crystals, you can simply collect them without the need for them to "do" anything else. Hobbies, such as collecting, have been shown to be beneficial for mental health, so if you have an interest in crystals, start a collection! The education in earth science and the camaraderie derived from meeting with other collectors will be far more satisfying and useful than the imagined energies your collection is said to provide.

And to be sure, when done improperly, crystal healing can actually be harmful. Using crystals as a substitute for proper medical care, spending too much money on all the "best" crystals, or using toxic minerals to "treat" your ailments are only a few of the ways that crystal-healing gurus can irresponsibly mislead their followers. Also, don't ingest, drink, or bathe with crystals, as you're potentially putting yourself at risk.

Remember, when reading an extraordinary claim about crystals, you must look for extraordinary evidence. Marketers of metaphysics would love for you to believe their claims, but we know better, and good scientific information is easy to find if you want to learn more about them.

Whatever your interest level in crystal healing, no matter if you are for or against it, never use it as a replacement for modern medicine. No matter how dissatisfied or untrusting of medical science you may be, voluntarily setting your healthcare back hundreds of years is an irrational and potentially dangerous reaction. Crystal healing and other folk remedies were once essential parts of medicine and community ritualism because there were no other options, not because they were superior to science. Today, crystal healing is simply an expertly marketed lifestyle product, providing no more connection to nature or our past than can be found simply by hiking a mountain trail or walking along the water's edge. The selling of well-formed crystals of rare minerals for healing is a prime example of the paradox of New Age metaphysics: why should these expensive specimens, divorced from their natural context, obtained through mining by unknown hands, be any more effective or important to you than a common stone found in nature, collected by your own hand?

Instead, get to know the rocks, minerals, and crystals near you, build a personal collection, and find friends and family to share them with. You'll likely end up with more improved health and happiness than crystal healing has to offer.

BIBLIOGRAPHY AND RECOMMENDED READING

Bates, Robert L., editor. *Dictionary of Geological Terms, 3rd Edition*. New York: Anchor Books, 1984.

Bishop, F. L., Yardley, L., & Lewith, G. T. (2010). Why consumers maintain complementary and alternative medicine use: a qualitative study. *The Journal of Alternative and Complementary Medicine*, 16(2), 175-182.

Bonewitz, Ronald Louis. *Smithsonian Rock and Gem*. New York: DK Publishing, 2005.

Chesteman, Charles W. *The Audubon Society Field Guide to North American Rocks and Minerals*. New York: Knopf, 1979.

Helffrich, G., Wood, B. The Earth's mantle. *Nature* 412, 501–507 (2001).

Hoenemeyer, T. W., Kaptchuk, T. J., Mehta, T. S., & Fontaine, K. R. (2018). Open-label placebo treatment for cancer-related fatigue: a randomized-controlled clinical trial. *Scientific Reports*, 8(1), 2784.

Johnsen, Ole. *Minerals of the World*. New Jersey: Princeton University Press, 2004.

Johnson, S. B., Park, H. S., Gross, C. P. & Yu, J. B. (2018). Use of alternative medicine and its impact on survival. *Journal of the National Cancer Institute*, 110, djx145.

Kaptchuk, T. J., Kelley, J. M., Conboy, L. A., Davis, R. B., Kerr, C. E., Jacobson, E. E., ... & Park, M. (2008). Components of placebo effect: Randomized controlled trial in patients with irritable bowel syndrome. *British Medical Journal*, 336, 999-1003.

Lack, C. W. & Rousseau, J. (2016). *Critical thinking, science and pseudoscience: Why we can't trust our brains*. Springer Publishing Company: New York, New York

Matute H, Yarritu I, Vadillo MA. (2011) Illusions of causality at the heart of pseudoscience. *British Journal of Psychology*, 102, 392-405. https://doi.org/10.1348/000712610X532210.

McClure, Tess. "Dark Crystals: the Brutal Reality behind a Booming Wellness Craze." *The Guardian*, September 17, 2019.

Moore, R., and McLean, S. *Folk Healing and Health Care Practices in Britain and Ireland: Stethoscopes, Wands, and Crystals*. Berghahn Books, 2010.

Mottana, Annibale, et al. *Simon and Schuster's Guide to Rocks and Minerals*. New York: Simon and Schuster, 1978.

National Center for Complementary and Integrative Health: https://nccih.nih.gov/health

Novella, S. (2012). *Why do people turn to alternative medicine*. Retrieved from https://sciencebasedmedicine.org

Pellant, Chris. *Rocks and Minerals*. New York: Dorling Kindersley Publishing, 2002.

Perceptions of Science in America (2018). American Academy of Arts and Sciences, Cambridge, Massachusetts.

Pough, Frederick H. *Rocks and Minerals*. Boston: Houghton Mifflin, 1988.

Robinson, George W. *Minerals*. New York: Simon & Schuster, 1994.

Sagan, Carl (1996). *The Demon-Haunted World: Science as a Candle in the Dark*. Random House: New York, New York.

Schmaltz R, & Lilienfeld SO. (2014). *Hauntings, homeopathy, and the Hopkinsville Goblins: Using pseudoscience to teach scientific thinking*. Frontiers in Psychology, 5, 336. https://doi.org/10.3389/fpsyg.2014.00336

GLOSSARY

Anecdote A personal account of an incident; considered unreliable or hearsay

Atomic bonds Enduring bonds (connections) between individual atoms made by sharing electrons

Atoms The smallest whole particles we currently understand; everything is composed of atoms. Atoms of different sizes have different properties; we call the differently sized atoms chemical elements. Each atom is comprised of a nucleus (containing protons and neutrons) surrounded by electrons.

Bias blind spot Dismissing or ignoring information that goes against firmly held beliefs

Causal illusion Perceiving a relationship or connection that does not exist

Causation When an action causes something to occur

Chemical elements The chemical "building blocks" of the universe; there are currently 118 known, though only 94 occur naturally. Each element is composed of atoms of a particular size.

Confirmation bias The tendency to seek only information and evidence that supports what you currently believe in while ignoring and dismissing information that runs counter to your beliefs

Core The innermost portion of the earth, consisting of a sphere of superheated metal

Correlation A connection or relationship (whether real or perceived) between two or more things

Crust The outermost layer of our planet; it is rigid and stiff, composed largely of hardened, cooled rock

Crystal structures The highly organized, symmetrical arrangements of molecules that make up crystals

Crystallize When a chemical compound hardens and takes on an organized structure determined by the atoms within it

Crystals The solidified forms taken on by minerals

Electrons Negatively charged atomic particles that orbit an atom's nucleus

Fact An explanation that is well-supported and deemed the most reasonable to believe out of all other proposed explanations

Falsifiability The ability for a statement to be contradicted or ruled out by evidence

Folk remedies Traditional medicines or treatments that are not prescribed by a doctor or necessarily supported by science

Hawthorne effect When a patient or study participant changes their behavior as a result of being observed by a doctor or researcher

Hydrothermal activity Water and steam naturally heated within the earth; the water can contain dissolved minerals that can be deposited as crystals inside cavities in rock

Hypothesis A proposed explanation for a particular observation, or an explanation that requires further testing

Igneous Rocks that form directly from the cooling and solidification of magma or lava

Jargon Specialized terminology used within professions or groups that is difficult for outsiders to understand

Lava Molten rock that has reached the earth's surface

Magma Molten rock within the earth

Magmatic crystallization When minerals within magma begin to cool, they crystallize and lock together to form a rock

Mantle   The largest layer within the earth; it consists of very hot, semi-solid rocks. The mantle is below the crust, but above the core.

Metamorphic rocks   Rocks formed when pre-existing rocks were subjected to heat and/or pressure, changing and reorganizing its constituent minerals

Metaphysical philosophy   A system of beliefs built around the idea that all things in the universe are connected via energy

Minerals   Inorganic chemical compounds formed when atoms of various elements crystallize (harden) together

Misattribution   Attributing an outcome or effect to the wrong cause

Molecule   A combination of two or more atoms of the same or differing elements (for example, sodium and chlorine) joined together; when identical molecules join together, it's called a chemical compound

Neutrons   Particles with a neutral electric charge within an atom's nucleus

Nucleus   The core of an atom. It contains protons and neutrons, and is surrounded by electrons

Occam's Razor   A philosophical principle that states that the simplest explanation is often the correct explanation.

Pattern detection   A mental tool we use to rapidly process information by learning to recognize regularities

Placebo   A mimic of a medication or therapy without the active drug or component of a treatment

Placebo effect   When a patient receiving a placebo shows or reports improvements despite not actually receiving the active component of the treatment

Precipitation When a solid substance is deposited by a solution, e.g., when mineral-rich water dries up, leaving the minerals behind

Protons Particles with a positive electric charge within an atom's nucleus

Pseudoscience A set of practices and beliefs that claim to be scientific but fail to meet the standards of the scientific method; pseudosciences cannot stand up to rigorous testing and are supported by unreliable evidence, such as anecdotal claims

Recrystallization When heat and pressure within the earth affect the structure of minerals, changing them into other minerals

Rocks A hard mass consisting of a mixture of minerals

Scientific law A well-supported hypothesis that has stood up to rigorous testing is considered a scientific law.

Scientific method A set of standards and practices within the scientific community for testing ideas and claims; gathering evidence in a systematic manner in order to develop a better understanding of a problem we wish to explain

Sedimentary Rocks formed when particles of other rocks, minerals, or organic matter accumulate and compact together to form a solid mass

Tectonic plates Huge sections of the earth's crust which "float" and move upon the soft, hot mantle

Theory An explanation for a set of related observations and events (e.g., germ theory, evolutionary theory)

ABOUT THE AUTHORS

Dan R. Lynch has written over twenty field guides, monographs, and kids' books on rocks and minerals, with a specialty in agates. He got his start at Agate City—the rock shop run by his parents, Bob and Nancy Lynch, in Two Harbors, Minnesota, where he learned the nuances of rock and mineral identification from a young age. After earning his degree in graphic design with emphasis on photography at the University of Minnesota Duluth, he combined all of his interests and expertise into a series of books that help new collectors "decode" the complexities of the earth sciences and mineral identification. Dan has made it his aim to communicate science honestly and factually; to dispel anti-science myths and misinformation with easy-to-understand language, an objective viewpoint, and photos that inspire wonder in the natural world. He currently lives in Madison, Wisconsin, where he works as an author, artist, and classical numismatist.

Crystallized magnetite—a favorite of the author

Julie A. Kirsch, PhD, research scientist and instructor in the field of health psychology, earned her doctorate from the University of Wisconsin–Madison. Building upon her master's degree in experimental psychology from Western Washington University, she has lectured and taught classes on research methods,

statistics, and the psychology of health and aging. With her multiple publications, including in *American Psychologist*, the flagship journal of the American Psychological Association, she has demonstrated her ability to communicate science effectively to a wide audience. And in drawing upon her expertise specializing in stress and its effects on mental and physical health, she combats misguided and unscientific practices prevalent in the health and wellness industry today. Her goal is to show her readers how to find more accurate and evidence-based approaches to attaining wellbeing and how to not let internet fads cloud their judgment. But science is a collaborative process, too; Julie enjoys expanding her horizons by working with other researchers and authors to further our understanding of our brains and behavior.